1

GODS, MAN, &WAR

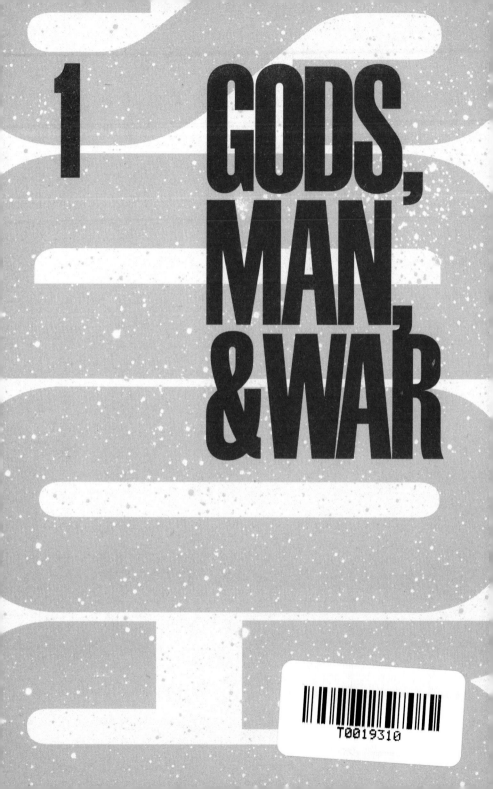

GODS

VOLUME 1

GODS, MAN, & WAR

An official
Sekret Machines
investigation
of the UFO
phenomenon

By
Tom DeLonge
with
Peter Levenda

Sekret Machines: Gods
Copyright © 2016 by Tom DeLonge

To The Stars, Inc.
1051 S. Coast Hwy 101 Suite B, Encinitas, CA 92024
ToTheStars.Media
To The Stars… and *Sekret Machines* are trademarks of *To The Stars, Inc.*

Managing Editor: Kari DeLonge
Copy Editor: Jeremy Townsend
Consulting: David Wilk
Book Design: Lamp Post

Manufactured in the United States of America

ISBN 978-1-943272-23-5 (Hard Cover trade)
ISBN 978-1-943272-27-3 (Hard Cover Limited Edition)
ISBN 978-1-943272-40-2 (Trade Paperback Edition)
ISBN 978-1-943272-28-0 (eBook)

Distributed worldwide by Simon & Schuster

*I would like to dedicate this
to anyone out there in the world
that knew there was a whole lot
more to our human story*

TOM'S ACKNOWLEDGMENTS:

I would like to thank my wife Jennifer, and my children, Ava and Jonas—For their support of my interests in all things weird.

I would also like to thank Peter Levenda for being such a brilliant researcher, an eloquent spokesperson for this important project, and a great friend. Through his words and experience, I was able to summarize this very important topic and eventually get us the ears of the US Government. Additional thanks to AJ Hartley for joining the fun, soaking in the esoteric knowledge, and bringing it all to life in a way for ordinary people to understand.

And last, to all my government advisors—

You have already begun to change the world through this project. People are waking up and thinking differently about themselves and the world they live in . . . all because of you.

PETER'S ACKNOWLEDGMENTS:

So many individuals have gone on before to research, write, speak and even at times risk their reputations if not their lives where the Phenomenon is concerned. It is inconceivable that more people do not take this enduring mystery more seriously, and it is our hope and intention that with this project more people will come to the conclusion that the missing piece of this puzzle is the missing piece of the human condition.

There is a group of about ten individuals we refer to as the "advisors." Most of these gentlemen cannot be identified here due to their standing in their respective agencies and departments. I ask the reader's patience with this omission, and state that I can vouch for the fact that they do, indeed, exist and that their contribution to this Project has been substantive and of great importance.

That said, there are some people I have to acknowledge by name, and thank, in this place:

Jacques Vallée, whose foreword to this book sets it squarely where it has to be. M. Vallée has been at the forefront of the post-World War II UFO Phenomenon for nearly the last fifty years or so. He and Dr. J. Allen Hynek were experts in the Phenomenon at a time when serious and respected researchers were in short supply; in addition, Vallée brought to the table his expertise in astronomy and computer science, making his contribution the beginning (and the model) of the multi-disciplinary approach to the problem that is so desperately required and which we are determined to implement as best we can. I am grateful for not only his foreword but for his continuing friendship and advice.

Whitley Strieber, who has been a friend since the 1990s and whose sincerity on the subject of UFOs and the abduction experience has been inspirational. Our voluminous correspondence over the last twenty years has yielded many insights and new avenues of approach to this very complex subject ranging from military and mind-control interpretations to the purely spiritual, and everything in between.

Thanks and acknowledgments must also go to those friends and colleagues with whom I have shared convivial meetings and discussions in the past on this subject, including Jim Marrs, Richard Dolan, Timothy Good and Hal Puthoff. For many in-depth radio and podcast interviews: Jimmy Church, Ian Punnett, the late Georgeanne Hughes, Dr. J. Ilias, Gordon White, Micah Hanks, Joseph Farrell ... well, the list goes on and on. These are the people who keep this discussion alive, especially at a time when many would like to forget all about it and move on to other subjects.

To Andrew J. Hartley, who is an incomparable raconteur as befits his dual role as best-selling novelist and king of Shakespeare studies (no allusion to Lear, Richard III, or the Scottish king is intended or implied!). Our conversations are usually wide-ranging and not without elements of wry humor and dollops of ironic asides. His fictional interpretation of the names, places and events described in this non-fiction series has provided a coherent narrative that has intrigued and even startled some of this Project's most seasoned advisors, so close it has come to being more than a novel and very nearly an exposé.

And, of course, to Tom DeLonge without whom there would be no such project and no such determined assault on the old, established theories of alien contact; and, of course, it has been Tom's unequaled access to high-ranking individuals in military, intelligence and corporate circles that has made this Project three-dimensional in scope if not four-dimensional.

Family and friends (and devoted readers) have been patient with me during the last roughly two years of work (so far) on this Project when I have been unavailable due to travel, research, and long hours of pondering the relationships between events, persons, and phenomena and the implications they suggest for a new understanding of the role of humanity in the world and for the extension of human consciousness into realms only dreamed of by scientists and shamans.

To all of these, and to those I cannot mention by name, my thanks and gratitude.

Peter Levenda

CONTENTS

FOREWORD by JACQUES VALLÉE

FOREWORD

When Tom DeLonge and Peter Levenda asked me to introduce this book, which constitutes a bold re-examination of the meaning and nature of unexplained aerial phenomena, I felt at the same time honored and challenged by the opportunity they offered, because the field has reached a crossroads.

The sightings have been ignored by academics, summarily brushed off by most politicians, censored by churches, classified by the military and ridiculed by the media. Yet the reports continue to come. The experience of the unknown has only deepened, raising unsettling questions about the intrusion of the uncanny in our shiny modern world of convenient machines and superficial entertainment.

First, questions about history. When the mythology about the origins of human cultures is confronted with environmental and biological fact, what truth emerges about the birth of humanity? What forces, what influences impinged upon its development? Were there gods on the earth in those hoary days (*in illo tempore*)? Or did we invent them to account for our spiritual need to touch the stars—a need born millennia ago, yet more urgent and powerful than ever in our own century?

The second question is about reality itself. It is even more scientifically fundamental and psychologically troubling.

Let's concede to the skeptics, for the sake of argument, that UFOs don't exist, as the experts of SETI and the consensus of Academe keep assuring us. Let us assume that reports of such objects are indeed inconsistent with anthropocentric ideas of physical reality *as the sophisticated edifice of modern science understands it.*

In a subtle twist of logic that would have delighted the philosophers of the Derrida school, this denial of the testimony compiled from hundreds of thousands of sightings all over the world suddenly enables us to accept them freely: What harm could there be in acknowledging these meaningless stories in the light of day? As the authors of this book argue brilliantly, once we agree that UFOs are impossible, nothing stops us from opening the files—even the secret ones: Indeed, "The UFO can be 'known' only by not asking what it is."

Which logically leads us to realize something else: If UFOs and physical reality are incompatible, maybe the time has come to re-negotiate physical reality. Because, as we all know, these impossible UFOs that don't exist are not going away.

Tom DeLonge and Peter Levenda, who have had privileged access to long-denied information, have re-opened the debate around these two questions. The persistence of the phenomenon in the full glory of its impossibility forces a fundamental re-examination of our history and our presence on earth, and it opens the heady prospect of a breakthrough beyond the confusion and ugly contradictions of modern physics.

Nothing is easy or simple once you take this step. As a reader of this book, if you agree with us to question old ideas and entrenched theories, you enter a world of dusty records, tentative interpretations and irreconcilable ideologies. As Jeffrey Kripal has noted, we dwell here "in a form of gnosis or forbidden knowledge well beyond reason and completely beyond belief."

To their credit, the authors are well aware that this ancient material is brittle. They bravely question the scholars' various hypotheses and the respectable traditions of established pieties, but they do not try to force upon us a new ideology of their own. Retaining the analytical sense that is critical in approaching this material, they display it before us in all its complexity. They invite you, as they have invited me, to join in a completely new phase of research, informed by the oldest and the most recent sources.

The "Sekret machines" have a message for us. The time has come to decipher it.

INTRODUCTION

In November of 2014, I received a strange communication from someone claiming to be Tom DeLonge, the front man for the famous band Blink-182. Although I had just turned sixty-four the previous month, and was therefore a bona fide Boomer, I was very much aware of Blink and my first reaction was "Yeah, right." It had to be some crackpot, looking for attention.

I got a lot of those over the years.

You see, my life has been somewhat strange in that I never seemed able to extricate myself from some of the weirdest connections imaginable for a kid who grew up in straitened circumstances in the Bronx—not to mention stints in Chicago and New England—and for a while I was only a handshake away from alleged JFK assassination conspirators David Ferrie and Jack Martin (portrayed by Joe Pesci and Jack Lemmon in the film *JFK*). I was only seventeen, a high school senior at Columbus High School in the Bronx, when a friend of mine and I gatecrashed the funeral for Bobby Kennedy at St. Patrick's Cathedral in June 1968.

That same month we made contact with a mysterious church that turned out to be some kind of anti-communist front with FBI affiliations, one that moreover included the aforementioned Ferrie and Martin as bishops.

How did we manage this, my friend and I? We simply created our own church out of whole cloth and incorporated it in the State of New York. It was my strategy of getting out of the Vietnam War. And it worked. Clergy can't be drafted.

So much transpired during the decade of 1968 to 1978, and I won't go into all of that here. So many connections. So many ramifications. In 1969, I began working for Spencer Memorial Church in Brooklyn Heights, where the pastor was sympathetic to the Weathermen and the Black Panthers who sometimes held meetings there at night. My job was to stay there on those nights and ensure that no one burned down the building. A short time after, the Weathermen did blow up a building in Greenwich Village, an explosion that I heard as it happened because I was on the phone to Presbyterian Church headquarters across the street from the doomed edifice when the blast took place.

In 1970, I started working for a company in the garment district of New York. It was owned by a guy named Willie Brandt. His son, Steven Brandt, had committed suicide in November 1969 at the Chelsea Hotel. Steve had been a close friend of Sharon Tate's and had been a witness at her wedding to Roman Polanski. After the Manson Family slaughtered the pregnant Sharon Tate and her friends in August 1969, Steve Brandt gave "voluminous" information to the police, and then attempted suicide. He did not succeed, returned to New York City, and finally did succeed that November.

In 1973, I began working for the Bendix Corporation, and was studying Mandarin Chinese. My office, Bendix International, was heavily involved in a series of political and intelligence situations at the time, as was revealed by the *New York Times* during Watergate.

In 1979, I was in Chile investigating the Nazi sanctuary Colonia Dignidad, and very nearly didn't return. I told that story in my book, *Unholy Alliance*.

In 1980, I began working for Bank Hapoalim, at their Rockefeller Center branch in Manhattan, sending coded traffic from the telex department to Tel Aviv.

In 1984, I began working for a small export company in Queens. That's where I met a colleague of E. Howard Hunt—the Watergate burglar and "plumber"—who worked alongside Hunt at the Mullen Corporation in D.C., and also in Singapore. It's where I first heard of the existence of something called AFIO: the Association of Former Intelligence Officers. I only lasted there three months.

In the summer of 1984, I began working for a tiny export firm about a block from the doomed World Trade Center. It was run by two Chinese, and I was the only employee. That job would begin my long and storied history in China trade, about which I would publish a book, which has gone through several printings (in China). It was also where I met Bernie Goetz, the Subway Shooter, at an electronics shop not far from the Trade Center.

And so it goes.

I did a massive amount of China trade during the period 1984–2001, with multi-million-dollar contracts (which was serious money in those days), and in 1994 I published my first book, *Unholy Alliance*. It would wind up changing my life. By this time, I was already forty-four years old.

▼   ▼   ▼

In 1996, I was working and living in Kuala Lumpur, Malaysia. I won't bore you with the details, but by then I had traveled to more than forty countries around the world for business. At the same time, I was compiling data for a book that I had conceived back in the 1970s about the weirdness of American history when seen through the lens of coincidence and synchronicity. By about 2002, I decided I would stop the research and actually start writing what would become the three volumes of *Sinister Forces*.

Back in the States in December 2003, when living in a Muslim majority country became a little hot after the invasion of Iraq that year, I finalized the draft of *Sinister Forces* and found a publisher.

As that was going on, I started corresponding with a professor at Florida International University in Miami. I had moved to Florida after Malaysia and was watching this guy on late-night cable drawing similarities between the Vedas and western philosophy and thought to myself, "This guy is good!" So, I wrote my first-ever fan mail. The upshot was I found myself enrolling at the university for a master's degree in religious studies, obtaining that in 2007.

In 2011, I was invited to speak at a conference on UFOs, "secret space programs" and "breakaway civilizations" in Amsterdam, along with Richard Dolan, Richard Hoagland, and Timothy Good, among others. I had no idea why they invited me to a conference that seemed to lean heavily into UFOs—hence Dolan and Good—but I prepared a talk on the American secret space program as I saw it: the connections between the infamous Maury Island and Kenneth Arnold affairs and the assassination of JFK. It was designed to be humorous as well as informative, but the loudest applause I received was when I made the off-hand statement that citizens should be able to keep their personal information private: that, in effect, they should be able to say to their governments, "No. You can't have that information. It's classified."

At any rate, that appearance sparked a small avalanche of requests to speak on various podcasts and radio interviews, including Coast to Coast, Midnight in the Desert, and Fade to Black, among others. While I had always been interested in UFOs and had maintained a voluminous correspondence with Whitley Strieber as far back as when I lived in Kuala Lumpur, it had never been my major focus. When I looked back over *Sinister Forces*, though, I realized I had devoted many pages as well as an entire chapter to the subject, but always from the point of view of American history, the military, synchronicity, etc. My perspective was more John Keel and Jacques Vallee than it was Hynek, Ruppelt, or even Good. Indeed, I had written about Hal

Puthoff, Jacques Vallee, and Whitley Strieber without ever imagining there would come a time when I would be sitting down with each of them.

It slowly dawned on me that UFOs might very well be the key to everything we are searching for, as human beings on this planet. They are the missing link in our consciousness, in our understanding of reality and of the parameters of time and space.

I had published books on esotericism, on Tantra, alchemy, and even on occultism, along with three books devoted to the subject of Nazism: especially their occult ideas, as well as the post-war Nazi underground (which I had personally investigated at Colonia Dignidad in Pinochet's Chile as well as in Argentina, Germany, and the United States).

It was while I was in this contemplative mood that I received the communication from Tom DeLonge.

▼   ▼   ▼

"Yeah, right." At least, that was my first reaction. I did some searching around, however, looking for some kind of verification that this guy was who he said he was, and discovered that Tom had been studying UFOs his whole life. Gradually, it sank in. The contact was real.

That was in November 2014.

One must understand the raw enthusiasm of this guy when it comes to UFOs. Enthusiasm mixed with unbridled optimism that we would get to the bottom of this. How would Tom succeed when no one else had succeeded these past nearly seventy years since Roswell? Well, for one thing, he would use the connections he had developed over the years as a celebrity. Celebrity opens doors, it is true, but you must knock on them just the right way.

Tom's approach was to begin from the assumption that the UFO Phenomenon—or what we would just start calling "the Phenomenon"—was real and proceed from there. Next, we would take a non-adversarial approach to the people Ufologists love to hate the most: the government, the military, the intelligence services, the industrialists. We would be polite, non-confrontational, and convince these individuals of our sincerity. Not only that, but we would have to demonstrate that we were not cranks or fanatics. We would have to show that we had a serious, thoughtful, carefully considered perspective on all this.

There was another angle, as well.

Tom had recruited a best-selling YA author and Shakespeare expert to write a fictionalized version of the story we wanted to tell. The reason for this was to be able to say things in fiction that we could not say in nonfiction for whatever reason, and to bring this type of information into the domain of younger people who did not grow up on Roswell, the Men in Black, or the paranoid, canned American history of *The X-Files*. Tom felt that the younger generation that grew up in the last twenty to thirty years would be the demographic that could effect the kind of change in government secrecy regarding UFOs that the older generation, who grew up during the Cold War, could not even imagine.

So, he began working with Andrew J. Hartley and me, right around the same time, in 2014. Hartley's task was to write a fiction trilogy, buttressed by the data we were accumulating on the Phenomenon, and my task was to write a nonfiction trilogy.

In 2015, we were going strong. We had spoken with such notable individuals as George Knapp, Bob Lazar, and a slew of others on the UFO conference circuit, but we also met most famously with John Podesta.

That was in June of 2015, in Washington, D.C.

By this time, Tom had been in regular contact with a small group of government insiders we started calling the "advisors." This group consisted of about ten individuals from various branches of the military and intelligence services as well as industry. I had completed the prologue to book one of the *Sekret Machines* trilogy, and Tom began to use it as a kind of "mission statement" for our project. It seemed to be well-received, for on the strength of that document (as well as Tom's exuberance and networking capability) we began to get positive feedback and introductions to some very highly placed individuals in the military. They started to take us seriously.

One of these led us to John Podesta.

Hillary Clinton had just announced that she was running for president, and Podesta was her campaign manager. He had worked at the Obama White House, where he had a reputation for being interested in the UFO phenomenon. Indeed, Mrs. Clinton herself eventually would go on record as being just as serious about the subject when asked about it by reporters and television personalities while on the campaign trail.

Twenty years prior to the 2015–2016 campaign season, however, Hillary Clinton had attended a gathering hosted by Laurance Rockefeller, himself passionately interested in UFOs. This was the Rockefeller Initiative, which occurred over the years 1993–1996 (Bill Clinton was president of the United States from 1992–2000) but which had its origins in 1992 immediately after the presidential election. Known formally as the "UFO Disclosure Initiative to the Clinton White House," it was designed to convince Clinton to declassify UFO documents, beginning with the Roswell crash documentation, dating back to July 1947.

While the Initiative had made inroads into the administration, not much was done to declassify important UFO records. The Clinton administration had declassified many documents pertaining to the post-war Nazi underground, the protection of Nazi war criminals by various governments including the United States (especially in the case of Klaus Barbie, the "Butcher of Lyon") and the flow of Nazi gold out of the occupied territories and into mysterious coffers abroad, including in Asia (all subjects I had written about in my books). It also had declassified many documents relating to the assassination of President John F. Kennedy (after the furor created by the Oliver Stone film, *JFK*). But UFO material largely had escaped attention.

In February 2015, Podesta tweeted that his biggest failure of the time he spent in the Obama White House as Chief of Staff was not getting the UFO files declassified.

We met with him four months later, and he explained the situation to us in general terms.

Basically, there is no single big box at the Pentagon (or anywhere else) that is marked "UFO" into which all UFO reports, analyses, videos, etc., are dumped. UFO data are spread across departments, branches of service, intelligence agencies, and the like and often in different places within those same organizations. Some UFO data are buried, dead on arrival. No one wants to deal with it. No one wants a UFO sighting on their permanent record. So, in order to declassify UFO files, one has to go step-by-step, agency by agency, and know precisely what to ask for and from whom to ask it. Then, in order for those documents to be declassified, everyone who is on the recipients list on each particular document has to sign off on it. So, if there is a UFO document that circulated among several different departments at the Air Force, for example, and then maybe also copied to the Army or the NSA, etc., each department has to sign off, and then the Army, the NSA, etc.

To make matters more complicated, Podesta allowed as the best way to get someone to read a memo was to classify it! In other words, people are much more likely to read a classified memo than an unclassified memo, contributing to a massive backlog of classified memos that should never have been classified in the first place, but which now have to go through the laborious declassification process anyway.

Basically, this was all a bureaucratic nightmare, but one that I have been at pains to discuss at the various UFO conventions around the country where I have been invited to speak.

A year later, and the "bureaucratic nightmare" would take on near-mythic proportions. In October 2016, there was the infamous Wikileaks dump of Podesta's email inbox, and it revealed that what Tom DeLonge had been saying about working with a high-level group of advisors in government was absolutely true. Tom's emails to Podesta were revealed, as were others mentioned by Tom, to the extent that our Sekret Machines project was blown. Tom was suddenly unreachable. His emails and phone went offline. The firestorm that accompanied these revelations in the UFO community was staggering. I became one of maybe only two persons publicly connected to Tom who was reachable (the other being A. J. Hartley) and there was nothing I could share with the general public nor with the radio and television interviewers who contacted me looking for some sign of life from Tom and from the project. The naysayers who formerly had believed that Tom was just making stuff up about his "advisors" were forced into a different channel of attack: that Tom's involvement with Podesta, with military brass, etc., was evidence that he was being used as a pawn by government insiders, the Men in Black, or the "deep state," depending on how you framed your particular conspiracy theory. Tom, who had jump-started the project alone with a few trusted colleagues, was now being characterized as a stooge who had been fronting for the "controllers" all this time. It was a joke to those of us who knew better, if not damned insulting considering all the work we had put into this project against impossible odds. My own published work over the past twenty-five years should demonstrate to anyone that I am nobody's fool and would never act as a "puppet" of the "deep state." I have been critical, openly, of government secrecy and of many United States government policies, particularly in foreign affairs.

Eventually, after a few months, the Sekret Machines project slowly came back online and a year after Wikileaks we had a different set of advisors. While those that I had met personally before Wikileaks were few, due to security reasons, I am happy to say that I have met most of those who populated the original board of To The Stars Academy of the Arts and Science because they were out in the open and listed transparently on the TTSA website, along with their previous affiliations with DOD, CIA, etc. No "mysterious whistleblowers" who can't be identified, but persons who were so high up in the government that they don't need whistles: persons who admit publicly that the UFO Phenomenon exists, that there is documentary evidence that it does, and that there was a secret department within the Pentagon that specialized in precisely that.

Just before the Wikileaks/Podesta flap, Tom and I met with a number of persons who would become prominent in the creation of To The Stars Academy. This was in Austin, Texas, in July of 2016. We spent a day with Hal Puthoff, Chris Mellon, and Jacques Vallee. It was my first time meeting any of these individuals. I hope I didn't embarrass myself as a fanboy, but I had written about Hal and Jacques in *Sinister Forces* and elsewhere, and I had critiqued one of Chris's relatives—Richard Mellon Scaife—as well. To sit at a conference table with all these guys was slightly mind-blowing, to be sure.

We were meeting to discuss the future of the Sekret Machines project, and to collate some of the material that Tom had accumulated over the course of his conversations with his advisers. Jacques Vallee was, and is, an iconic figure in the UFO universe and he does not suffer fools gladly, if at all. His perspective on the advisers and what they were telling us was sober, thoughtful, and reflected decades of personal experience in dealing with the military and the intelligence community, as well as with scads of Ufologists, experiencers, and members of bizarre sects and societies in the United States and around the world. Chris Mellon is the ultimate insider, both due to his family connections as well as his long government career that includes stints as deputy assistant secretary of defense in both the Clinton and Bush administrations.

Hal Puthoff has a career that began long before the early days of the remote viewing projects at SRI International in the 1970s and his partnership with Russell Targ in investigating the paranormal. Like Jacques Vallee, he has degrees in the sciences and technology. Between the two of them, they have met some of the most famous names in Ufology as well as the paranormal.

So that meeting in Austin was certainly stimulating on so many levels. Hal, Jacques, and Chris all have security clearances so they could not always discuss what they knew in front of Tom and me, but thankfully we rarely ran into a topic that they could not clarify. One of the major issues was the idea that the US government knows much more about UFOs than they are telling us. That they have more *data* than they have revealed so far is certain and would be demonstrated by the release of the Tic Tac video in December 2017. However, UFO datum does not lend itself easily to interpretation and certain knowledge. In other words, they may not *know* more about the UFO Phenomenon now than we do; what they do have is information that defies analysis. In other words, we're back to square one.

The problem may be the approach taken by those with control over the data. Either the UFO is a threat to national security, or it is not. Unfortunately, the relevant organs of the US government are not in agreement about this. One of Chris Mellon's criticisms of the government is the fact that it does not take the possible threat seriously even though UFOs have buzzed nuclear facilities and military bases, not only in the United States but in the former Soviet Union as well. Paradoxically, one of the popular criticisms of To The Stars Academy is that they take the threat potential *too* seriously.

The second problem is a shadow of the first. Since the gear heads at the Pentagon cannot understand how the UFO is designed and constructed, or what its form of propulsion might be, they throw up their hands in frustration and concern themselves with other things. To be sure, there are urgent threats abroad in the land from international terrorism to cyber warfare to political and military flashpoints around the globe. UFOs wind up on the back burner. Why take the UFO seriously as a national security threat when it hasn't dropped any bombs—yet—or decommissioned our missile bases—at least, not permanently—or otherwise committed a hostile act? This, in spite of the fact that the UFO Phenomenon has demonstrated an alarming interest in our Navy and our Air Force operations.

This fact, this cat-and-mouse game that the Phenomenon seems to be playing with our armed forces, raises a host of other questions that the Pentagon is not equipped to answer. These are existential questions, problems of psychology and culture, of language and consciousness, and that is where our general, official approach to the Phenomenon has failed us.

One take away from the meeting was that the machinery (the UFOs themselves) is real (observable, powerful, caught on video tape and electronic sensors), but that its impact on consciousness—first the direct effect on the experiencers and then the indirect effect on the public at large—is just as real, and understood even less. If a foreign country were behind the UFO Phenomenon, it would be considered a form of psychological warfare. In fact, it's the kind of stuff *we* used to do during the Cold War: in Vietnam, Cuba, Africa, and elsewhere. Look up Operation Phoenix, as an example.

In addition, there is a possible genetic effect upon those who have had a close encounter, which should raise another set of alarms. The possibility of "genetic warfare" was broached as early as 1970 and by 1997 William Cohen, Secretary of Defense under President Clinton, was discussing "ethnic specific" pathogens and other "ethnic weapons."

The idea that there might be a connection between Nazi scientists and the UFO Phenomenon was raised by Tom, on and off, since the project began and again during the Austin meeting. Shortly before the day of the meeting there had been news reports of a recently discovered tunnel complex in Austria that had served as part of the Nazi secret weapons research program. It is well-known that, after the British raid on the German V1 and V2 labs at Peenemunde during World War II that the bulk of its equipment and personnel had been moved to a vast, hastily created tunnel complex called Mittelwerk near Nordhausen in Germany. This discovery of another complex in Austria was startling, to say the least. We remembered that many Nazi scientists were brought to the US under Operation Paperclip, but that there were similar programs run by other countries was not as well known. Was it possible that there was a weapons or aviation research program in Latin America or Africa or Asia that was responsible for the late 1940s sightings of UFOs? If so, that would have complicated matters entirely!

At the time, the consensus in Austin was that a Nazi flying saucer program was not a credible thesis for a variety of reasons, most of them logistical. The possibility would stick in the back of my mind, however, as my own research into the Reich over the previous decades raised some serious questions about the feasibility of a shadow Nazi technology enterprise, as we will see in the third volume of the *Sekret Machines* series.

After a very long day of meetings, lunch at the conference table, and a collegial dinner afterward, I got back to my hotel room and started poring over my notes. There was a lot to work with, but the overriding take-away for me was the fact that I had observed the most knowledge-able individuals in the United States on the subject of the Phenomenon discussing probabilities and possibilities (to the extent they were able to do so) in solemn, very sober terms. There was not a person present who did not take this subject extremely seriously. At one point, during a discussion on a particularly sensitive topic, there was a sudden pause in the conversation. Glances were exchanged among Hal, Chris, and Jacques. Hal finally said, "We would have to continue this discussion in a SCIF . . ." and the topic was dropped. Tom and I just looked at each other and shrugged.

That October saw the Wikileaks dump of the Podesta emails, and we would go dark. Our network of advisers rolled up, some of them understandably pissed. But the first volumes of the *Sekret Machines* trilogies—both the fiction and the nonfiction—had been published. We were getting our message out regardless of Julian Assange, Wikileaks, Pizzagate delusions, or Russian skullduggery. People slowly began to understand what we were trying to do, and moreover they realized—albeit reluctantly in some cases—that Tom was deadly serious about his goal. What confused some critics is our insistence that both a fictional and a nonfictional approach to the material is necessary. We knew we had to transcend some of the fixed categories that had put blinders over the eyes of the more dedicated research-ers, but which more importantly had confused the general public whose only exposure to the subject of the UFO was in terms of fiction, in books and in film. This confabulation of the real with the imaginary created a false narrative about the Phenomenon, so much so that it became impossible to disentangle the threads. Serious scientists and academics avoided the subject due to the damage it would cause to their credibility if they came within arm's reach of the UFO. This opprobrium extended even to the military itself, which was the target of conspiracy theories about Area 51, crashed saucers, and alien bodies. While some officers of high rank and lofty security clearances might have known more about the reality behind UFO sightings—or at least took them seriously based on what evidence they did possess—the average noncom and commissioned officers were encouraged to ridicule the entire subject.

Using the approach of "informed fiction" gave us the opportunity to present this material in a narrative fashion that would engage the average reader, many of them coming across this type of information for the first time. The *Sekret Machines* novels were not written with the average Ufologist in mind, but rather for a generation that had not yet been poisoned by the cynicism and skepticism that had so far bedeviled any serious discussion of the Phenomenon. We wanted to stimulate discourse on this important subject and encourage a younger crowd to come up with their own ideas and perspectives, to get them to "buy in" to the fact that the UFO Phenomenon is real but that it remains so far inexplicable. We knew that renewed interest in STEM and especially STEAM subjects in a population that consistently tests lower than other industrial nations in math and science would provide a bright and educated base for coming up with new solutions to this ancient conundrum, by encouraging new generations to think outside the box to seek alternate theories and methods. It was thought to make science and math exciting by demonstrating their application to the sexiest scientific, technical, and cultural problem of modern times: the UFO Phenomenon.

Tom suspected that his fans would not be willing to wade through a heavy academic treatise to get to that point. His fans were, after all, his first point of contact with the public and they were people who loved his music. He shared their enthusiasm, their culture, their aesthetic. He was fascinated by the UFO Phenomenon and he felt certain some of them would feel the same. The novels by A. J. Hartley satisfies this need to speak to a larger demographic than the one represented by the UFO convention circuit: the older, mostly white, politically conservative people who attended UFO conventions or listened to a lot of talk radio. Instead, we wanted to direct our efforts towards a new demographic largely unaware of the complexity and profundity of the Phenomenon except as a punch line. This would not be an attempt to resurrect the old clichés about flying saucers and little green men, but to create or inspire a new narrative involving technology, history, and culture. This new narrative had to be framed, presented as a series of possibilities, given a context that was relatable. The novels provided that context, a starting place for discussion. (Take the well-known example of the popularity of the Indiana Jones movies responsible for creating a new generation of archaeologists.)

At the same time, the "heavy academic treatise" approach had its good points, too. We wanted to pose questions and stimulate thinking along different lines and to show how the tremendous advances in science, technology, artificial intelligence, archaeology, anthropology, and the humanities could contribute to a new understanding of the Phenomenon. The nonfiction approach represented by *Sekret Machines: Gods, Man and War* was intended as a companion series to the novels, introducing readers to three specific knowledge areas with relevance to the Phenomenon: religion and spirituality; science and technology; and war, conflict, and sovereignty with a focus on the history of the last one hundred years.

This was not an "ancient astronauts" approach. We were not interested in interpreting every unknown as "it must be aliens." That might be good for entertainment, but our sights were set higher than that. We had to keep the hocus-pocus to a minimum and not insult the reader's intelligence. We wanted to provoke, but not confound. So, the nonfiction volumes (including, of course, this one) are heavier and denser than the novels as we crafted what is essentially a legal brief or an academic dissertation with the goal of making a case for the serious, mainstream, very public and eventually very respectable study of the UFO.

When we came online again, in October of 2017, there were new players and a new organization: the To The Stars Academy of Arts and Science. In addition to Hal Puthoff and Chris Mellon, there was Jim Semivan, formerly of CIA, and Steve Justice, formerly of Skunk Works. There was also a guy called Lue Elizondo who, as the world would soon discover, was formerly head of the Pentagon's super-secret (and thus far completely unknown to the world) UFO investigation body called AATIP.

It was time for us all to throw away our tinfoil hats. The UFO was going mainstream.

▼   ▼   ▼

By now, you all know what happened. The *New York Times* carried the story, co-authored by Leslie Kean, on December 17, 2017: There was a secret group inside the Pentagon that was tasked with monitoring UFO sightings and trying, against all odds, to make sense of them. That was AATIP, the Advanced Aerospace Threat Identification Program, and it was run—until the previous October—by Luis Elizondo. That is when the floodgates opened and every major news outlet (print, digital, and television) started carrying the story. UFOs were real. The military knew they were real. They had them on video, for crying out loud.

Since then, besides the *Times*, the *Washington Post*, *Scientific American*, *Popular Mechanics* even, everyone was covering this story and taking it seriously. No tales about little green men or crackpot eyewitnesses. No laugh-tracks. People weren't laughing anymore. This was now a mainstream story and it changed the landscape forever. People could talk openly about UFOs and the "Phenomenon" without risking their reputations or being considered gullible or lunatic. Technology had caught up with the eyewitnesses. There were video recordings: UFOs caught on FLIR cameras from fighter-jet cockpits. Radar operators freaking out. Computer systems aboard ships and planes struggling to make sense of it. And as this is going to press, there are more and more stories coming out of the US military, more video footage, more testimony from sailors and airmen.

More evidence. More data. More proof.

What Tom DeLonge started in 2014 came to fruition in 2017: a mere three years later. Now UFOs are respectable. Talk of alien races, inhabited planets, the possibility of communicating with extraterrestrial species, traveling inter-dimensionally, etc., etc., are subjects that are seeping into the national and international conversation in scientific magazines and newsletters, as well as in the popular press. This is a more sophisticated, more intelligent dialogue than anyone expected just a few years ago to be coming from the government, the military, and academia.

This book, co-authored with Tom and published in 2016, along with *Sekret Machines: Chasing Shadows*, co-authored by Tom and A. J. Hartley, started it all. You're in for a mind-expanding (if not mind-blowing) experience, reading both these books side by side. If you've read them before, they reward a second read in light of everything that has happened since they were first published five years ago.

Go back and take another look, and then look up. Things have changed. The skies are getting crowded.

The machines are still "sekret" . . . but not for long.

Peter Levenda
2021

PROLOGUE

The Stone Age

The sound from heaven was like that of thunder, but it was a clear night. They stood frozen where they walked beneath the dense tropical canopy at the rumbling growl from the sky.

The oppressive heat was a physical presence they ignored. Water dripped from the leaves of the dense foliage all around them. This was their land, a land they knew intimately. They fished, they foraged for herbs and tubers, they built huts made of bamboo and grasses. They lived, they ate, they had children. But this . . . this was not from their land. It was not even from their earth.

They became statues, wide-eyed and trembling in the moonless dark, transfixed by what they heard; afraid of what they would see. Their chief was summoned, but there was no need. He already had been alerted by the sound, the insistent thrum that descended upon them from the sky.

Then the heavens opened and the night was full of light and fire.

Above the tree line, above the roof of palms and ferns, they could see the sun shining impossibly at that darkest hour. The jungle around them shivered, the ground vibrating with the steps of some unseen being. Most fled, to the doubtful security of straw walls and old habits, but the distance they covered in an hour was only a second's work to the secret machines of the gods in the sky.

The ones that remained heard inhuman voices booming from above their heads. Lights played all around them, penetrating the branches of trees, the puddles of rank water at their feet, frightening the snakes, the rodents, the birds into taking flight. What power on earth or in the sky could turn night into day?

They crawled on their bellies, seeking the camouflage of weeds and grasses, and crept along the jungle floor towards the unholy din, the clamoring of demons, the ceaseless clattering like the shaking of dried peas in a gourd only so much louder, so many more gourds, so many countless numbers of peas. But these few had to know. They had to see the source of this light and this terrible sound. They were the elders of the People. They were the only ones who could understand the meaning of the sounds, of the lights, of the horror.

A kilometer further down a hill—a mound sacred to their fathers, for reasons no one remembered—they came upon a clearing and their hearts leapt into their throats. What they saw was impossible. What they saw no man had ever seen. What they saw had no words in the language of the People to describe.

Beings, clothed in light, descended from the skies. There were spheres turning in all directions. There were faces, like the faces of the People, shining from every direction. There was a canoe—a kind of canoe, a vessel, like a gourd—rising up from the ground and was the source of the insistent throbbing noise that had aroused them from their slumber hours ago.

The elders kept watch. In their minds, they tried on different words—like hats—for their images to describe what they saw. So they could tell the People when they returned. They were witnessing the arrival of beings with tremendous powers, beings who controlled light and sound and could fly through the air. They heard the voices of these beings— huge voices, voices that could carry through the air like the drums of the People—but they understood not a word. They saw symbols, and they had no word for symbols. They were pictures but they were not images of anything they had ever seen. The elders knew, without expressing it in words, that what they were experiencing was a moment of initiation. It was a spiritual event, a crossing over into another existence.

The lights, like little suns, like giant stars, illuminated the night.

▼   ▼   ▼

By morning, the images became clearer. There was more color. More activity. The elders could see beings that looked like People, and they were very busy. A huge building of some kind—but a building that could move all over the ground—was the source of many wrapped packages. These packages would be distributed to various beings, who then took them to other places, other buildings, sacred shrines or gravesites.

The elders quietly discussed whether they should approach these beings. Whether it was safe. Whether they would be welcomed. But before they could make a decision, there was another terrifying sound from the sky.

A sudden, blaring, shrieking noise caused the elders to drop to the ground, prostrate, in the presence of the most powerful, most unearthly event in their ancient history. Another "building" came flying down from the sky, and came to the ground some distance away. It crawled over the earth until it was close enough to see clearly. And from its stomach more packages were removed.

The elders took careful note of the design of the temple and its broad avenue. They noticed the lights. They noticed a high place, made of wood. At the top of that high place, the beings seemed to speak directly to the Father of the Sky.

The packages were filled mostly with things the People did not understand, but with some things they did. Some seemed to be food, for the beings ate from them. Others seemed to be implements of some kind. Clothing. Water. The elders smacked their lips at the sight of all that bounty.

They returned to their village. They told stories of what they had seen. Supernatural beings from the sky. Flying devices. Light. Sound. And the many, many packages sent from the Beings.

The People asked them many questions, over and over again. Finally, it became clear: The elders had been initiated into the mysteries of the Light Beings. They had become "illuminated." (Only people who lived in darkness could appreciate the divinity of the Light.) They knew what to do. They knew how to summon these Light Beings so that they, too, could receive the gifts from heaven.

They made ceremonial clothing in imitation of what the Beings wore. They made implements in the same design as those of the Beings, magic devices to communicate with the Spirit in the Sky, magic devices to fly, magic devices to see at long distance, magic devices that made terrifying sounds. They found artifacts on the ground when the Beings finally left to return to their villages in the heavens, and they kept them as sacred relics of power. These machines were kept apart from the People and only revealed on sacred days. The machines contained power and knowledge, and access to that power and knowledge was the privilege of those who had seen the Beings firsthand. There were no words to describe all that had been seen, no vocabulary available to people living in the Stone Age, so the essence of these machines remained secret, wrapped around with ritual language and arcane ceremony that made sense only to those who had seen.

The People built a broad avenue in the jungle near their village. They erected a high tower like the one they had seen. They stationed their elders on top of that tower to scour the heavens for a sign that the Beings were returning.

And they created a prayer: "Spirit of the Sky, Remember!"

▼   ▼   ▼

The People lived in the Stone Age, but in the midst of the twentieth century. In 1942. They lived in the South Pacific, on islands that had been contested by Japanese and British, Australian and American forces during the Second World War. They had never seen aircraft before, or motorized vehicles. They had never seen Japanese or European men. And the effect of all of this was the creation of a new religion based on gods who descended from the skies, bringing wisdom and knowledge, Coca-Cola and hot dogs, machine guns and medicine.

This religion is called by anthropologists and journalists a "cargo cult." It exists to this day.

Quite possibly it has always existed. Everywhere. Quite possibly since the beginning of recorded history.

June 1947
Wright Field, Ohio

The balding, middle-aged, fifty-two-year-old German with the autocratic air and the slight mustache was not diminished, even after two years in British custody as a war criminal. Instead, he stood on the tarmac and breathed in the Midwestern air as a free man. More than a thousand of his compatriots and comrades had already made the journey from Peenemünde to America, many of whom had worked for him during the war. It would be nice to see the old fighters once again.

Major-General Walter Dornberger was to be put in charge of guided missile research, which was appropriate considering he had developed the V-1 "buzzbomb" as well as the V-2: the world's first long-range guided missile. More than three thousand V-2s were launched during the last months of the war, against London and Antwerp, killing thousands of civilians.

He heard that his colleague, SS Major Wernher von Braun, had also made the trip to the States and was headed to White Sands Proving Ground. The Americans would not send Dornberger there, or to Fort Bliss, Texas, where the rest of his team was forming up. They were cautious, thinking that it would be unsafe to put the general back in command of his old division.

Didn't the Americans realize that he did not have to appear physically before his men to ensure their allegiance, their loyalty? They had not become suddenly patriotic citizens of the United States overnight, with nothing more than six dollars a day in wages and what passed for food in the miserable American wasteland, abandoning their ideals and their political and spiritual beliefs for sliced bread and a cold cot. No. They were still his. They were still von Braun's.

▼   ▼   ▼

In three years, Dornberger would wind up as an executive with Bell Aerospace but at that moment, in June of 1947, a thousand miles away in the New Mexico desert, something had just crashed. Weather balloon, Project Mogul, or—as the Air Force initially reported—a flying disk; whatever it was, it had just ushered in a new reality.

A new normal.

Terre Haute, Indiana
Federal Prison
February 14, 1950

After eight years doing time for sedition, a crazed mystic is released from federal prison and returns to his home in Noblesville, Indiana. His theories about race, spirituality, UFOs and alien presence on the Earth have culminated in an ideology he calls "Soulcraft."

He was in prison as the leader of a pro-Nazi organization called the Silver Legion that demonstrated unconditional support of Hitler and resistance to the government of the United States, in particular the administration of President Franklin D. Roosevelt. His name was William Dudley Pelley, and he was an intimate of such infamous American Nazis as Father Coughlin, Fritz Kuhn of the German-American Bund, and others. He was also a friend of George Hunt Williamson: the confidant of George Adamski, another crazed mystic with UFO associations. At one time, before the war began, Pelley's was the largest pro-Nazi organization in the United States. Now it had taken the same turn as Adamski: to the stars.

Pelley's book *Star Guests: Design for Mortality,* was published in 1950, the year he was released from prison. Like

Andrija Puharich and his guests at a mysterious Maine estate three years later, Pelley was in communication with an alien intelligence. He had been since the evening of October 28, 1928, at nine p.m., at an apartment in New York City, where a trance medium delivered the first of many alien messages. Like Puharich and his circle, he was receiving guidance from these "star guests."

And like Puharich and his guests, this guidance was intended for political, spiritual and cultural change in the United States specifically and in the world as a whole.

July 1952
Washington, DC

The United States witnesses a major overflight of a squadron of UFOs over the nation's capital. This event is recorded on film as well as by numerous eyewitnesses, to the extent that the US Air Force is pressured into giving the largest press conference in American history to date to debunk the sightings. The overflights occur on two successive weekends in July. Had this been done by a terrestrial power it would have been considered a gross provocation and an incitement to war. Instead, it was white-washed by the Air Force in official statements. Their own investigation of UFO sightings—known as Project Blue Book—had begun only a few months earlier, in March of 1952.

Privately, the US government began to explore social and cultural counter-measures in terms of propaganda and psychological warfare, such as had been used on civilian populations in Europe only a few years previously during World War II, in order to denigrate further sightings of UFOs by civilians. The Robertson Panel was established in January of 1953 for that purpose. They meet for a total of twelve hours. The panel concluded that there was no need to devote time and energy to the investigation of UFOs and that, in fact, civilian UFO study groups should be put under surveillance as possible sources of subversive activity.

The policy backfired.

The Round Table Foundation
Near Augusta, Maine
June 27, 1953, 12:15 a.m.
The night of the full moon

Less than a year after the DC sightings, a gentleman from Puna, India, is sitting in the center of a circle in a house in the woods, channeling beings from the sky. It is a dark night in the middle of a remote New England location, far from the lights of cities and towns. He is fingering a string of beads. He is in a trance.

Then, a voice, identifying itself only by the letter R:

Tonight we want to create Brahmins in this world . . .

From out of nowhere, a lump of cotton string appears before the assembled participants in this unlikely séance. This is known as an *apport*, something that seems to materialize out of thin air. The medium gave each of the participants one of these strings. They are what Brahmins wear after a ceremony has been completed. The participants are told they *are* Brahmins.

Six months earlier, these same beings from the sky had communicated through the Indian medium a command from Above:

> *We are Nine Principles and Forces . . . We propose to work with you . . . Peace is not warlessness. Peace is the integral fruitage of personality. We have designed to utilize you . . .*

Now, six months later, nine participants have gathered to make contact with these nine beings. To become "utilized." To become Brahmins.

The participants are known to history. The names of these Brahmins were recorded for posterity. They are:

Henry Jackson and his wife, Georgianna Jackson. Henry Jackson would become an important and well-respected hospital administrator in California for many years.

Alice Bouverie, nee Astor. Wealthy, entitled, and with mystical preoccupations.

Carl Betz, later a Hollywood actor, famous as the TV husband of Donna Reed.

Vonnie Beck, of whom not much is known.

Marcella Du Pont, a friend of the author H.L. Mencken, who will go on to become an important figure in another channeled movement as translator of *The Urantia Book* into French.

Andrija Puharich, the leader of the group. A medical officer with the US Army at Edgewood Arsenal. A scientist and occultist. The man who wanted to weaponize ESP. The man who would discover Uri Geller, the Israeli psychic.

Arthur Young, the inventor of the Bell Helicopter. An engineer and occultist.

His wife, Ruth Forbes Young. Socialite with a distinguished pedigree. Best friend of Mary Bancroft: mistress of Allen Dulles of the CIA. Mother of Michael Paine, an employee of Bell Aerospace in Dallas. And mother-in-law to Ruth Paine of Irving, Texas, the woman who got Lee Harvey Oswald his job at the Texas School Book Depository.

▼   ▼   ▼

A Forbes, a Du Pont, an Astor. American royalty. American wealth.

Henry Jackson and Andrija Puharich, medical men with hard science backgrounds.

Arthur Young, an engineer with a hard science background.

Ruth Young, a mystic with intelligence agency connections and a hand-shake away from a presidential assassin.

In contact with beings who claim they are aboard a spacecraft in near-Earth orbit.

Beings who are giving them *instructions*.

▼   ▼   ▼

This project is not intended to convince you of the reality of UFOs. If you need convincing—after all of the data that has been presented by sober, sane members of world governments, including that of the United States, as well as by military observers around the world who have gone on record concerning alien contact—then there is nothing here for you. We will not burden this narrative with the usual accumulation of dozens or even hundreds of sightings over the years. These have been covered in books, articles, television programs, and websites and are easily accessible to everyone.

Instead, this project is predicated on the understanding—not the belief, the understanding—that the UFO phenomenon is real and that there has been contact between human beings and non-human beings since the beginning of recorded history with results that can be characterized as alternately positive and edifying, and dangerous and terrifying.

This project is predicated on the understanding that there has been an explosion of this contact in the last seventy years, and that disclosure of the nature of this contact is not to be expected from the United States government or from its military or political leaders but that it is nonetheless happening every day, in every country on Earth, to the average man, woman and child.

Therefore, the time has finally come to stop apologizing for the recognition that the UFO phenomenon exists and that it has serious implications in all areas of human endeavor. It is time to stop characterizing this recognition as a "belief system," on a par with Santa Claus and the Easter Bunny. The tension between the authority figures who deny that this phenomenon represents anything real, and those average citizens who realize that it *is* real based on either their own experience or on the records of sightings by responsible individuals, has led to a dangerous state of affairs in which the disconnect between the state and its citizens has led not only to a crisis of confidence in government but also to a critical state of inaction in the face of what could be either a threat of universal proportions or an opportunity for dramatic growth forever lost.

▼   ▼   ▼

Here is how this project is designed:

> . . . for both intellectual and mystical reasons, I am unable to draw any sharp distinctions between the "real," the "religious," and the "fictional."
> — Jeffrey J. Kripal, *Authors of the Impossible:*
> *The Paranormal and the Sacred,*
> (Chicago: University of Chicago Press, 2010), p. 34

Using the above quotation from Dr. Kripal's important work on the range of the paranormal experience—including especially those experiences that fall under the rubric of UFO—as our own framework, *Sekret Machines* intends to demonstrate that by merging fictional and nonfictional approaches, including mass media and social media in a variety of strategies, something analogous to the "truth" may be discovered about the foremost challenge to global culture in the twenty-first century. This challenge we have decided to identify as "the Phenomenon."

We use this designation in order to emphasize certain salient characteristics of what has been described variously as UFO (Unidentified Flying Objects) or UAP (Unidentified Aerial Phenomena). From the vast literature available on the subject, we have come to the inescapable conclusion that what we are dealing with is a *phenomenon* that transcends historical, scientific, and religious contexts. It involves not only actual sightings—including but not limited to still photographs, film, radar traces, etc., but at times physical contact with the Earth as well as with human beings on the Earth. It involves various forms of communication; violations of what we understand to be physical laws; impossible forms of propulsion; psychological disorientation in observers; physical trauma to observers; religious visions; anomalies of all types; paranormal events including telepathy, telekinesis, etc., and a resulting confusion in our political, military, and industrial sectors. This is a phenomenon that has been experienced since the earliest days of recorded history, virtually without any significant deviations from era to era. Thus, to characterize it as UFO or UAP or "flying saucers," etc., is woefully inadequate. Instead, we wish to subsume all of the above characteristics and experiences under the single rubric of the Phenomenon, and it is in this manner that we shall refer to it in the remainder of this work. It is a Phenomenon that shows no distinctions, as Kripal has noted, between the real, the religious, and the fictional. For that reason, this project consists of three works of fiction, three works of nonfiction, and associated documentary and feature film treatments as well as a comprehensive social media platform. It is hoped that by enjoying and experiencing the full range of what we offer that a deeper understanding not only of the Phenomenon but of the nature of what we call reality will become clear. It is also hoped that this approach will make it easier for those in other disciplines—such as medicine, physics, biology, chemistry, astronomy, astrophysics, engineering, information technology, psychology, sociology, anthropology, and religious studies (to name but a few)—to openly discuss and research the Phenomenon, and to create excitement and creativity in those within the educational system, both students and teachers, so that new advances in science, technology, and philosophy will result.

As the world speaks of the convergence of technologies it has not addressed the necessary psychological and philosophical mindsets that will be required by the new technological environment. We are still thinking like seventeenth-century citizens even as we use twenty-first-century technology. Our worldview is limited even as our horizons have been extended: it seems we can only see so far without falling back on old prejudices, badly informed cynicism, violence against our neighbors and against our planet, and a sense of the futility of human action to effect change. Even though Galileo proved to us that the Earth revolves around the sun, and not the other way around, we are still behaving as if the world we experience every day is the only one there is. We are constrained by our worldview: literally, by our *view* of the *world*. It is an image, a construct, a fantasy, that has outlived its usefulness.

We dare, in this project, to present an alternative.

Thus, this project is designed to inspire nothing less than a cultural revolution in human consciousness. Once it has been accepted that alien contact is real, has occurred, and is occurring, then an overhaul of our religious, cultural, political, scientific, and military preconceptions is not only required but is inevitable. Thought leaders in each of these fields are urged to re-examine what they already know and to reframe their knowledge in light of alien contact. To do otherwise is to whistle in the dark.

▼   ▼   ▼

We will, over the course of this and the following volumes, attempt to provide the framework for this re-examination. We will begin with the earliest accounts of alien contact—those contained in some of the core texts of the world's religions—and go on from there to describe the effect of this contact on other areas of human life. This is not an attempt at an "ancient aliens" type of approach; we are not interested in recycling that material, or in attempting to prove that every perceived anomaly on the planet is the direct result of alien interference or guidance. Instead, we are proposing that the history of human civilization over the past ten thousand years or more is nothing other than a

Cargo Cult. We will recreate the initial contact(s) as best as we can, with the data available to us, and then re-imagine our collective human history from that point on.

We call this initial contact the *ur-punkt*, the origin point: the Alien Genesis.

▼   ▼   ▼

As we move forward, a few concepts will become important. One of these involves the very use of the word "alien." The term has become a pejorative, a word used to demonize a people, or a belief system, or to ridicule the very idea of contact with beings from off-planet. To use the term "alien" is to telegraph a certain mindset where it doesn't exist and to devalue the dialogue in the process: people who speak of aliens are people who wear tinfoil hats or hear voices in their heads. Unfortunately, we do not have an adequate alternative to this term and for the time being will continue to use it until a better one comes along.

Another concept is that of disclosure. We do not believe that waiting for disclosure to come from the government is a wise position to take. It implies that the government—any government—controls access to alien contact. That is manifestly absurd. While we understand we cannot fathom the motivations or intentions of an alien intelligence—except by deducing such intentions through the evidence that we collect—we are certain that a human government would not be able to control communications, appearances, or access to alien contact the way they can over human contact, and that human governments have been demonstrably unable to do so. Further, there is no advantage to be had by any government in disclosing the reality of alien contact, because to do so would be to admit a degree of impotence in the face of their constituents that would challenge all social institutions.

Another concept that requires reframing is that of re-engineering. There have been books and articles written concerning the possession of alien technology and the efforts by governments to re-engineer this technology in order to discover its function and operation and to develop technologies of human origin based on discoveries made during the re-engineering process. While this may indeed be the case—and there is evidence to suggest that these secret machines have been developed for decades, if not longer, but that *they may not be what we think they are*—we believe it is more important to consider that it may not be alien technology that is being re-engineered.

The re-engineering may be taking place much closer to home than any of us can imagine.

It is virtually impossible to describe this project without sounding hyperbolic or deluded, a problem that is more an artifact of the current state of knowledge of what it means to be a human being than it is of the subject matter itself. We are still far away from a profound comprehension of who we are, of understanding our place on our own planet (much less in the cosmos at large), and this colors our appreciation of forms of knowledge that do not adhere to a geocentric view of human history. Galileo moved our center of gravity from the Earth to the sun in the seventeenth century; but our culture has yet to catch up to him.

It should be noted that only one of us has seen UFOs. While Tom has, Peter has not. However, Peter is convinced of the reality of the UFO phenomenon based on examination of the evidence. He does not believe that a scientific approach to the problem of UFOs should require personal contact or observation. If it did, it would be the only scientific phenomenon to do so. We routinely accept the existence of phenomena we personally do not observe, such as bacteria and sub-atomic particles, because the evidence is overwhelming. The same is true of UFOs. The resistance of some members of the scientific, political and military communities to acceptance of this phenomenon is based on a desire to withhold this information from the civilian population for reasons of "national security." The resistance of other members of these same communities to acceptance of the reality of this phenomenon is not based on science or on a logical assessment of the evidence, but on a cultural bias.

The intention of this project is to challenge all of that and to offer an alternative approach, a modern twenty-first century reboot of the old ideas, and to probe into the true nature of the mystery behind the Sekret Machines.

A Note on Sekret

We are using this unusual spelling of secret quite deliberately. For all of its street associations it also helps us to distinguish what we mean by secret from the usual interpretations.

The spelling "sekret" is also an Eastern European allusion, and much of what we will be discussing has Cold War associations including Soviet "sekret" machines.

In some Eastern European and Scandinavian languages, as well as in German, *sekret* means "secretion." This has special resonance for our thesis when we consider the nature of the Phenomenon as a "secreted" machine. This will be elaborated upon in the second and third volumes of this series.

Finally, we also want to emphasize the alternative meaning of "secret," derived from the Greek, which is "mystical."

A shrine of the Cargo Cult of "John Frum" on Tanna Island, Vanuatu, in the 1960s. Notice that the cross is red, perhaps in emulation of a Red Cross symbol seen on the "mysterious" sky ships? John Frum is believed by his followers to be a World War II era serviceman who will come back to the islands to bring wealth and prosperity (cargo) from the sky.

Walter Dornberger (left) and Wernher von Braun (center). These were two of the many Nazis brought to the United States under Operation Paperclip after the war. Von Braun would be sent to Huntsville, Alabama and Dornberger would be sent to Wright Field near Dayton, Ohio in June, 1947: the same month as the Roswell crash debris was sent to Wright Field.

The Wanted poster for William Dudley Pelley, American Nazi who was imprisoned in the USA during World War II for sedition, and who was a believer in UFOs and benign "Star Guests" after making contact with them during a séance in New York City in 1928.

The helicopter inventor Arthur Young. Young, credited with the design of the Bell Helicopter, left that company at the end of World War II to devote himself to a study of the paranormal. His wife, Ruth Forbes Paine, was an heiress and best friend of Mary Bancroft: the mistress of CIA director Allen Dulles. She was also the mother-in-law of Ruth Paine, who was the woman who got Lee Harvey Oswald his job at the Texas School Book Depository. Both Arthur Young and his wife, Ruth, were members of "the Nine": a group of nine influential Americans – including an Astor and a DuPont – who believed they were in contact with inhabitants of a UFO during a séance in 1953.

Ava Alice Muriel Astor, socialite and daughter of John Jacob Astor (who died aboard the *Titanic*). She was one of "the Nine" along with Marcella DuPont, Arthur Young, and Ruth Forbes.

SEKRET MACHINES

GODS

VOLUME 1

GODS, MAN, & WAR

ALIEN GENESIS

Since the very earliest times of mankind, there has existed a particular mental attitude on the part of man as regards the existence of a thought supposed to be superior to his own: this is the religious attitude. Until now, human thinking has never been applied to a category of thought supposed to be super-human other than in a religious context . . . The particular difficulty of Ufological research is, consequently, the difficulty of applying oneself to a super-human phenomenology merely with the methods of science and excluding all mysticism.

— Aimé Michel[1]

PROFESSOR THORKILD JACOBSEN—SCHOLAR OF ancient religions—once wrote that "basic to all religion . . . is, we believe, a unique experience of confrontation with power not of this world."[2]

Hans Jonas, the authority on Gnosticism, wrote concerning the God of the Gnostics: "the alien taken absolutely is the wholly transcendent, the 'beyond,' and an eminent attribute of God."[3] The Gnostics saw God as an alien, as a being from elsewhere who sojourned on Earth for a time.

That is not to say that either of these two authorities were speaking of "alien astronauts"; they were not. They were, however, coming to grips with the concept of the divine and how that experience was understood by ancient peoples as both alien and terrifying. In Jacobsen's case, he was writing about ancient Babylon in the third millennium BCE; in the work of Jonas, we are talking about a sect that flourished in the first few centuries of this era. Although they represent cultures separated by thousands of years from each other, the reaction to the divine experience was similar: the alien, the strange, the terrifying.

Gnosticism is the word used to describe a loose affiliation of mystical groups that flourished in Egypt and the Middle East in the first few centuries of the common era and that survived in some parts of the Middle East to the present day, incorporating some Islamic elements along the way. The word *gnosis* means "knowledge" and it is used here to describe a movement that arose out of Jewish, Christian, and later Islamic thought that claimed special insight into the mystical or hidden aspects of their respective faiths. In particular, the groups we consider Gnostic today mostly have their origins in early Christianity although there was a heavy Jewish component as well (Christianity began, after all, as a Jewish sect). Gnostics felt that the story told in the Torah—the first five books of what Christians call the Bible or "the Book"—is an exoteric version of a deeper, esoteric truth and that the Jews and Christians had misinterpreted or misunderstood the texts.

For instance, a key element of much Gnostic thought is the doctrine of the Demiurge. In this retelling of Genesis, the creation of the world is said not to be that of God but of a subordinate creature known as the Demiurge, a term that comes from Platonism and which refers to the Creator but considered to be responsible only for the material universe. To the Gnostics, the Demiurge is opposed to spirituality. In the Garden of Eden, according to this idea, the Serpent who tempted Eve was the Supreme Being; the Demiurge was the one who warned Adam and Eve against eating the fruit of the Tree of Knowledge. Famously, the Serpent advised Eve that if she and Adam ate of the forbidden fruit they would become "as gods" (Genesis 3:4).

Thus we have the roots of a very different kind of creation story, one in which human beings are striving for godhood against the wishes of a jealous Creator. This situation implies a great degree of alienation from the world, from matter, from what we may call "consensus reality": the reality upon which we all agree is the only true reality. The God of the Gnostics is "alien to that of the universe, which it neither created nor governs and to which it is the complete antithesis."[4] Human beings, according to this view, were created by lesser creatures who themselves have no direct knowledge of the true God and who, indeed, are obstacles to divine revelation.

This Gnostic concept should resonate with those who believe in the contemporary theories of humans as a hybrid race, designed only for labor in the service of alien beings, and they may be surprised to learn that this concept gained a considerable number of followers almost two thousand years before our current crop of "ancient alien theorists." Confusion may arise in that both the Creators and the true God are aliens in this view and there is, indeed, a spiritual dualism implied that is anathema to orthodox Jewish, Christian and Muslim believers. However, to those who theorize that there may be multiple categories of beings who interfere with or who influence humanity at various times, this purely Gnostic concept may prove interesting if not valuable.

Gnostic theory was quite detailed and insistent on this concept and claimed that the entire universe was a prison with the Earth its deepest, darkest dungeon.[5] Around the Earth in an array of concentric spheres were the realms of the planets and the stars. In some cases, these spheres were limited to seven; in others, they extended outward from the Earth for a total of 365 spheres (certainly a symbolic number, and reminiscent not only of the number of days in a solar year but also the number of years the prophet Enoch lived on the Earth before he was taken bodily into heaven). While this arrangement seems like pure Platonic theory it also has elements in common with the earliest recorded civilization in the world, that of ancient Sumer.

Perhaps the foremost proponent of an ancient astronaut theory involving Sumer is the late Zecharia Sitchin (1920–2010). A former London journalist and native of Azerbaijan who relocated to New York, Sitchin taught himself Sumerian in an effort to decipher the ancient cuneiform texts of that civilization. His work has been criticized by scholars of Sumerian, Akkadian and the civilizations that centered in and around Babylon. Sitchin's translations of key texts as well as his decipherment of some cylinder seals was idiosyncratic, to say the least, but that did not stop his works from becoming worldwide bestsellers, translated into forty-five languages. Less a scholar than an intuitive—one of those writer-mediums of which Professor Jeffrey Kripal writes[6]—Sitchin nevertheless touched on some specific weaknesses and speculations of current astronomical theory, introducing the idea that at some point in the distant past there was a planet or planets in our solar system that exploded, a theory put forth by the late astronomer Dr. Thomas Van Flandern to account for both the asteroid strike that killed the dinosaurs sixty-five million years ago as well as the existence of a civilization on Mars.[7] Sitchin's main theory, however, that human beings are a hybrid race created as slaves for extraterrestrial rulers, has no actual basis in the Sumerian texts themselves, nor is the existence of his "twelfth planet" evidenced by the Sumerian cylinder seals even though he has insisted that such evidence exists. Yet, something about this story—spread over more than a dozen books— touched a nerve in the general population. It seemed to explain something, to fill in a blank in the human record that has not yet been filled by evolution or astrophysics or genetics. As wild as Sitchin's theory may be, there is an element of something in the midst of all the mistranslated Sumerian, the mistaken Aramaic, and the unsupported astronomy that resonates with those who understand that there is a continuing presence in the world that cannot be explained by science at its present level of understanding, and who do not feel they can wait for the scientists to catch up. So they gravitate to works like those of Zecharia

Sitchin and Erich Von Däniken and others, and to the cable shows hosted by Giorgio Tsoukalos and David Hatcher Childress, because at some level they know there is truth there.

Somewhere.

There is general distrust of authority in the world, whether it is of governments or of academia, of the military and the scientists. So conspiracy theory that used to focus on political assassinations has now expanded to include metaphysical speculations as well. The crossover from political conspiracy theory to ancient astronaut theories, Holy Grail romances, and Ufology has created an underground that is part art and part literature, part science and part magic. Part politics and part mysticism.

> Events in my life caused me to start questioning my goals and the correctness of everything I had learned. In matters of religion, medicine, biology, physics, and other fields, I came to discover that reality differed seriously from what I had been taught.
>
> —Thomas van Flandern[8]

As Samuel Noah Kramer wrote, "history begins at Sumer"[9] and with the beginning of recorded history we have the beginning of paranoia and the suspicion that all is not what it seems. While the truth may not be as Sitchin conceived it, there is still a strange and persistent element of terror in the cuneiform texts out of Sumer that may reflect concerns among these ancient peoples that are eerily similar to our own.

One of the core texts of ancient Sumer is their creation epic, the *Enuma Elish*. Discovered in 1849 by the great archaeologist Austen Henry Layard (1817–1894) at Nineveh (near Mosul in present-day Iraq), it is a text of about one thousand lines in cuneiform and records the "origin story" of the Sumerians. While Sitchin and other ancient astronaut theorists spin complex tales of aliens and hybrids based on imperfect readings of various texts, the *Enuma Elish* is actually quite straightforward.

According to this document, the cosmos in its earliest form before the Earth and the sky were created was composed of two gods: Abzu (the "far water") and Tiamat (whose name means "mother of life" but who also represents Chaos), representing fresh water and salt water, respectively. Within Tiamat's body rest the other gods, including their leader Enki (in Akkadian, Ea) who will become an important deity to the Sumerians. According to the story, the gods make so much noise that Tiamat desires to kill them. There is a revolt, however, and Marduk becomes the leader of the noisy gods and manages to kill Tiamat, separating her body into two with which he creates the world: one half for the Earth, the other half for the sky. He then kills her general, son and consort, Kingu, and from Kingu's blood and Marduk's own breath creates the human race.

There is no academic controversy over the authenticity of this text, and very little controversy over its translation. Scholars agree that this is the Sumerian creation epic and, as such, represents the earliest recorded creation story in human history. That should not be taken as evidence that it is somehow "true" in any kind of scientific sense in the way we understand science, but it is an indication of a deeply held idea about how the universe was created and ideas about the origin of the human race. This is a story in which human beings were created from the blood of one god—an older god, representing the "establishment" of Abzu and Tiamat—and the breath of another god, representing an antagonist, the leader of a rebellion, in the great cosmic drama. In other words, human beings are seen as a hybrid race formed from two separate and warring divine beings: a hybrid that can be said to be a composite of matter and spirit, of blood and breath.

When we look back on this now from the perspective of our modern context, we tend not to assign too much importance to this epic. We have a tendency to treat narratives like this as fairy tales, superstitious nonsense, the blind grasping of primitive minds towards a reason for the existence of matter, spirit and life who fall upon a supernatural explanation that really explains nothing but which serves as a means for subverting any further investigation as rebellious or even treasonous. If we use the system of cultural anthropology, however, we would be forced to acknowledge that the *Enuma Elish* represents something that—in the context of its time and place—was accepted as true in some way and which reflected an understanding of human origins that was prevalent in Sumer more than five thousand years ago.

In more recent times, scholarship such as that represented by *Hamlet's Mill* and the heavily criticized works of L. A. Waddell among others, offered a different approach to the problem. In the case of *Hamlet's Mill*,[10] the theory was offered that much of what passes today for ancient mythology was actually an attempt to describe astronomical phenomena using the code words of gods, demons, serpent monsters, and the like. In the case of Waddell, there was the attempt to show linguistic similarities between widely divergent ethnic and cultural groups (the Sumerians, the Chinese, the Egyptians) in an effort to demonstrate a common racial origin.[11] The more popular works by Sitchin and von Däniken build on these premises, and on the highly controversial claims by Robert Temple[12] and others that even the African tribe known as the Dogon were aware of astronomical phenomena about which they could not possibly have had any direct experience unless it had been brought to them by another civilization possessing either advanced telescopes, or which had been traveling back and forth to the stars (in either case this would not be a terrestrial civilization).

No matter which source one consults, however, from the dusty cuneiform tablets of the Sumerians to the popular texts of the ancient astronaut theorists, one thing seems consistent: religion as we know it has its origins—either in reality or in fantasy—in the stars, and there seems to be an agreement that humans have a divine (or at least astral) origin. The Tantric texts of ancient India and the Egyptian narratives both describe the appearance of the cosmos as the result of a sexual act by the gods. The Bible specifically identifies the creation of Adam as an act of God who mixed mud with his own breath (a retelling of the Marduk story) to create a being in his own "image and likeness." That breath has a divine origin itself is attested not only in the Bible and in the Kabbalistic works of the Jews but also in the Indian Tantras and in the yogic practice of *pranayama*, as well as in the books of the European alchemists.[13]

This commonality of themes *should be* unexpected if not impossible. After all, human beings roamed the Earth and lived in different climates, different ecological environments, developed tools at different points in their evolution, spoke different languages, and were racially and ethnically distinct. From a postmodern perspective it would be foolish to insist that there is any kind of common or "ur-mythos" that humans from entirely different backgrounds would share. Yet it is obvious that human beings from entirely different cultures, of different races and ethnicities, share some ideas in common such as rituals surrounding birth and death and puberty: events that happen within all human societies and for which various "myths" are composed, recited and enacted. These rituals provide a social function and contribute to the coherence of the community that celebrates them, as well as suggesting a larger context for the shared human experience.

Keeping that in mind, we should remember that rituals and myths that pertain specifically to supernatural beings and their interactions with human beings are also fundamental to many societies around the world and have been since the beginning of recorded history if not earlier. Creation myths usually refer to supernatural events and supernatural beings; it is rare to encounter an explanation for the existence of human beings that does not involve the actions of gods. As humans we are obsessed with origin stories as if knowing where we came from is relevant to who we are today and what we will become tomorrow.

Origin Stories

There is something deeply satisfying in knowing, for instance, the origin story of Superman or Batman, or any of the superheroes that have populated western culture in the last eighty or ninety years. That these are fictional characters is irrelevant to the allure they have, and we cannot stop making movies about these comic book personalities or retelling their adventures in issue after issue. They represent, in a sense, a kind of popular religion: supernatural beings who come to the rescue of humans and who represent moral and ethical values. There is nothing new in this analysis; it has been done many times in the past.[14] Yet, if we draw some similarities between these characters and the religions of ancient Sumer, Babylon and Egypt (as examples) we will see that not much has changed since then except perhaps for one salient feature not found in the earlier epics but essential to many of the new ones: The secret society.

Superman came to Earth on a space ship from an exploding star. Already we are in Zecharia Sitchin territory.

While Superman's special abilities do not exist on his home planet, the Earth's environment interacts with his genetic heritage and gives him super powers. Superman is an alien, a visitor, an extraterrestrial biological entity, or EBE. Since Superman was created in 1933 and did not become a DC Comic until 1938, however, the connection between Superman and the UFO phenomenon was not a possibility at the time. Superman was instead a creature of the Second World War and the rise of Nazi Germany and its secret weapons. He was, quite literally, an *Übermensch* in the Nietzschean mold, but one working for the Allies rather than the Axis powers. He was also an illegal alien.

This "illegal alien" had godlike powers and characteristics. He fought existential threats in the form of supervillains like Lex Luthor; and he had a consort in the form of Lois Lane. He was a god who did not enslave human beings but who came to save them (mostly from themselves). He thus had a great deal in common with ideas about divinity that we find everywhere from India to Indiana, but with a special attribute: no one really knew who he was.

Clark Kent: "Catfish"?

Superman's identity was secret. In order to live among human beings it was necessary for him to assume a false identity. He became Clark Kent, the "mild-mannered" newspaper reporter for the *Daily Planet* (yet another allusion to extraterrestrial themes). He walked among the citizens of Metropolis like anyone else, even more so. With his horn-rimmed glasses, fedora, business suit and retiring manner, he appeared more like a clerk in an insurance company than a divine being who could see through walls, fly through the air, speed around the world faster than light, and lift entire buildings with his bare hands.

The association of enormous powers with a secret identity has a precedent in the history of religions but it is a relatively late one. While there have been secret initiation rituals in many cultures—the rites of Eleusis and of Mithra come to mind—it was not until the *Rosicrucian Manifestos* of the seventeenth century that the idea of a secret group of individuals with extraordinary powers became common currency. It was a secret society whose members also had a high moral and ethical character, as revealed in their *Manifestos*. There was also an "origin story" concerning their founder, Christian Rosenkreutz, who had traveled to the mysterious East (most likely Yemen) where he obtained the knowledge necessary for the making of gold and the *elixir vitae*: a medicine that would cure any sickness. He returned to Europe and founded his mystical fraternity for the good of humanity. He eventually died and, according to the Manifestos, was buried in a secret location, the Vault of the Adepts.

At about that same time we experienced the rise of Freemasonry, another secret society whose members' special powers resided more in their sense of shared community and intellectual freedom than in changing lead into gold or administering the elixir of life. Their initiation rituals refer to an ancient ancestor or prototype, These—the

Rosicrucians and the Freemasons—were the Clark Kents of their age.

It may be that the Enlightenment idea of a secret society whose members have special knowledge about the workings of the world became a lens through which later writers would understand the ancient texts of Sumer and Egypt. Did these civilizations possess secret knowledge about the origins of the world and especially of human beings, knowledge that was suppressed by later cultures and institutions such as the Catholic Church, only to surface in the occult rituals and secret teachings of the Rosicrucians and Freemasons? If we are to believe Sitchin and von Däniken and the other "ancient astronaut" theorists this indeed would seem to be the case. And the popularity of Superman (and the rather darker figure of that other secret superhero, Batman) would be evidence that—on some level, some *unconscious* level—people understood this to be true.

The Cult of the Reptilian

In ancient Sumer there were individuals who worked marvels in secret. We don't know quite what to call them, but they were feared by the general population. Terms like "witch" or "sorcerer" would be used to describe these individuals: persons who worked magic against the innocent.[15] We do not know if these individuals were members of a society of like-minded "ritual specialists" or if they worked as solitaries, but from the cuneiform texts we understand that they were feared, not admired. The well-known Akkadian

Maqlu text (first millennium BCE) provides lengthy incantations and rituals designed to protect against this form of magic and its practitioners, who are known as "sorcerers" or "witches." This type of personality would appear in the Bible in various places: as the "witch" of Endor who could raise the dead (1 Samuel 28: 3-25), or in the famous admonition, so freely interpreted and abused, "thou shalt not suffer a witch to live" (Exodus 22:18). More to the point, however, is the reference to "let them curse it who curse the day, who are skillful to rouse Leviathan" (Job 3:8).

By the time Genesis was composed, Sumer already had been invaded and the Sumerian people were fast disappearing, although their language remained as a sacred tongue among the Babylonians right up to the first century of the Christian Era. Motifs from the Sumerian Creation Epic survived throughout the ancient Near East in the story of a massive serpent, a sea monster that was identified with the earliest moments of Creation. In the above-mentioned passage from Job, the "rousing" of Leviathan is associated with curses and with those who work dark magic, the "witches" of the Maqlu text. The female Leviathan had a mate in the male Behemoth according to Jewish tradition. While Leviathan was a sea-dweller, Behemoth roamed the dry land.

Reptilian monsters who represent Chaos and who threaten human life are to be found in different cultures around the world. The human-reptile struggle is replicated from the war between Marduk and Tiamat to St. George and the Dragon. While the serpent was considered to be the natural enemy of human beings, there were those who worshipped them or who tried to control their powers for use against other human beings. There were even those, like the Gnostics, who considered the Serpent in the Garden of Eden to be the real god, the Alien God.

According to Gnostic sources, it would be Adam who would make common cause with the Alien God. If we are seeking some kind of moral equivalency in these narratives, we will not be satisfied. This is less a question of good versus evil than it is "Alien vs. Predator," with human beings occupying the middle ground between them. The Serpent in the Garden was an interloper, a stranger, who nonetheless knew that the Demiurge who created the world was passing himself off as the "One True God" when in reality he was a subservient figure who desired to keep humanity weak and in the throes of pure materialism. "Ignorance is bliss," so the saying goes, and the Demiurge wanted to keep humans in a state of ignorance. The Serpent wanted to awaken humans from their sleep, their lack of awareness and their state of spiritual oblivion. Thus according to the Gnostics, Adam did not "fall" from grace; that is, he did not fall asleep but instead fell awake.

In one Gnostic narrative current among a sect known as the Mandaeans, the Alien God—referred to as such, and also as the Stranger or the Messenger—came to the Earth from space. He descended through the orbits of the planets to finally arrive on Earth with the goal of redeeming humanity through knowledge.[16] There was opposition to this from the gods of the planets who wanted to thwart the Alien God in his mission of liberation.

Without delving too deeply into the story, the essential elements are plain: there are human beings on Earth and they live in a state of unconsciousness. There are two alien forces on the planet, struggling with each other for supremacy. One force wishes to keep humanity asleep and unaware; the other force wishes to give humanity knowledge so that it may free itself of the clutches of the planetary forces, the Archons or Aeons, and escape the prison that is the Earth. The Alien God took the form of the Serpent in the Garden of Eden and advised Eve that if she and Adam ate of the forbidden fruit that they would become "as gods."

Thus you had those for whom the Serpent—including Leviathan, the sea monster of the Bible—represented pure evil, the Devil and Satan: these are the followers of the orthodox forms of the Abrahamic religions. Then there were those, like the Gnostics, who saw this demonization of the Serpent as the creation of what we would call today the "mainstream media." The Gnostics claimed to know the real story, and in a sense their version of Genesis could be called an early form of conspiracy theory. Reflexively, contemporary conspiracy theory—a phenomenon known worldwide today—could be considered a modern form of Gnosticism. If we include Ufology—and many authors who had specialized in investigative journalism, political history and conspiracy theory moved to Ufology or even occult theories later in their careers, and vice versa[17]—then we have the perfect environment for the development of something that transcends both Ufology and conspiracy theory, a theory of reality that is as grounded in science and politics on the one hand and in mysticism and psychology on the other: a grand unified theory of consciousness which embraces the world of phenomena of every type, phenomena of what we might call a non-rational or "spiritual" type as well as those of more fundamental material forms, seeing the one as the extension of the other in a multiverse, or parallel universe, or in some version of hyperspace such as those mathematically represented by superstring or supergravity theories: an environment that would require a specially trained consciousness to experience and within which to maneuver.

Indeed, one can interpret the Demiurge as the patron saint of science: it represents the material world and all that is in it. The Serpent represents a hidden force within the material world, penetrating it, making contact with human beings when the grasp of materiality is at its weakest. This idea has been taken up time and again as serpents in many cultures represent spiritual power, rejuvenation (the shedding of the snake's skin), the coils of Kundalini in Indian Tantra and yoga, and the twin serpents on the caduceus of Hermes and Mercury, among many other symbolic representations.

Recent research into the Genesis story and especially attempts by modern archaeologists to locate the physical site of the Garden of Eden has led to some interesting revelations.[18] New information has led some scholars to believe that there was a race in the Near Eastern region that predated the arrival of the Sumerians. This race employed words which were later adopted by the Sumerians but which were not originally part of their language. The names of Adam and of Eden itself are both borrowed—so the theory goes—from the language of an older race, called by some Proto-Euphratian or Ubaidian, just as Sumerian would become the sacred language of the Akkadians and the Babylonians. In this schema, the word Eden simply means "fertile plain," and the word Adam did not refer to a person but to a "settlement on the plain." Thus, in a new interpretation of Genesis, the people of Adam had to flee their fertile plain. This was due to rising sea levels in the Persian Gulf which were obliterating the plain and submerging two of the four rivers that were said to be part of the Garden of Eden. LANDSAT images have discovered the missing two rivers, below the waters of the Gulf south of what is now Iraq.

By the time the Jews had incorporated this story into Genesis, presumably after the Babylonian captivity from where this rendition is inherited, many of the original ideas were truncated and the words in some cases lost their original meanings. Other Sumerian texts, such as those describing the Great Flood, would make their way in abbreviated or altered form into Genesis as well. To some, this would detract from the work of Sitchin and others who see the Sumerians possessing secret knowledge from the stars instead of survival stories from the seas. However, this need not be the case.

The more research that is done on the stories of the Bible with confirming data from other Near Eastern documents (as well as modern technological advances in ground-penetrating radar, satellite imaging, etc.) the more one realizes that these are not fantasies of superstitious minds but reflect attempts to describe real events that occurred in real life to real people. There will be those who will see in the latest research confirmation of their theories that the Bible stories were based on older Sumerian and Proto-Sumerian "myths," which themselves were little more than memories of a better life before climate change transformed the contours of the Gulf region and forced hunter-gatherers to join with the agriculturalists, thereby losing their freedom in the process. Now forced to live in established settlements, growing food and raising livestock, the dwellers off the land became chained to a life that promised material wealth and safety in exchange for giving up a life of adventure and challenge. It was, in a way, a defiance of a God who had always provided before and who now found himself being replaced by the farmer, the builder, the accountant, and the insurance salesman.

That is only half of the story, however. Taken out of context one can come away with the idea that the ancient peoples did not believe in spiritual forces at all, but simply used a kind of coding system to write down the important events of their history. The problem with this approach is that these same writings include detailed descriptions of other worlds as well as other-worldly inhabitants. Eden, after all, was always on the Earth. It was *Heaven on Earth*:

a simulacrum of another place, a point on the globe where Heaven, however briefly, made contact. Just as a presumably real place called Eden, the "fertile plain," became an earthly paradise in Genesis, real events transpired to give the Sumerians (and, later, other Near Eastern peoples such as the Akkadians and the Jews) a complex mythologem of astral beings descending to the Earth to impart wisdom or to wreak havoc. The emotional content of these stories constitute evidence that we are not dealing with normal disasters and the violent acts of ordinary humans but with something else entirely. Genesis places the very human history of the pre-Sumerian "Eden" within a much larger context of spiritual forces and the spectacle of human confusion in the face of the seemingly irrational behavior of non-human antagonists. It would be a mistake to imply that religion—or spirituality, or whatever term you wish to use at this point—did not exist in Eden or in early Sumer because Genesis is nothing more than a history book.

▼ ▼ ▼

We need not look further than the Old Testament of the Christian Bible to come across numerous instances where confrontation with the divine means abject terror. The God of the Israelites is a vengeful, wrathful deity who destroys entire cities, empowers armies, demands bloody sacrifice. This is a power "not of this world," a power from "beyond."

In the western world, the relationship between humans and gods has been problematic from the beginning. This is especially true of the monotheistic religions whose deities are jealous, angry, punitive, and harsh. They do not share any human attributes, such as those enjoyed by the gods in polytheistic religions, except those that demand obedience and sacrifice. To the Abrahamic gods, humanity is a pathetic mess that needs straightening out. Or else.

From the perspective of an outside observer this relationship seems to be based on some specific previous experience of both the gods and the human beings. At some point there was a breach of trust, a betrayal, or some kind of disconnect between the two types of existence. The Book of Genesis describes it as the result of humanity disobeying God's command not to eat of the Tree of Knowledge in the Garden of Eden. While largely understood as a metaphor, what is important to realize is that this story describes an act of will on the part of human beings that caused a rupture in the god-human relationship. This implies there was a time, *in illo tempore,* when human beings lived in a blissful state, free of worry, work, and death. This seems to be taken for granted by most of the world's religions, who almost uniformly characterize our contemporary worldly existence as the result of a fall from grace, a degeneration of the human condition. How strange, then, that science tells us that human beings are the result of *evolution*, not *devolution*.

Evolution, of course, does not imply that human beings are getting *better*, only that they are changing in response to the demands of survival and the stresses of an uncaring environment. In fact, we are told quite clearly that evolution favors only the continued survival of the species in general, not of individuals within that species. We, as humans, are only vehicles for the propagation of our genes. Our genes do not care about our individual identities, our histories, our first kiss, our last gasp; their only concern is that we procreate. Nothing else is relevant. Consciousness itself—especially self-awareness, the experience of an individual identity—may be superfluous. After all, the instinct to procreate is just that, an instinct. It does not require much in the way of consciousness, just a functioning set of reproductive organs.

Thus, the biblical (and other) stories of creation should be viewed from that perspective. "Be fruitful and multiply" (Genesis 1:27) was the command given to us by the God not of the Jews but of the Genes.[19] Procreating does nothing for the parents; it does everything for the genes. We are genetically disposed to having children, caring for them, and then dying off once we have done our job.

Then what is the point of religion? What is the point of a perspective of reality based on individual salvation as opposed to the collective, if our only reason for existing is the survival of the entire species? Are religious laws designed for the same reason, to enforce a system of morality, an ethical code, that favors the group over the individual? The Ten Commandments seem to reflect this point of view, beginning with laws that uphold the authority of God—"Thou shalt have no other gods before me"—and then proceeding to a series of regulations that have as their sole concern the survival of the group: from "thou shalt not kill" to "thou shalt not steal," etc. The moral code is a social code; there is no context for individual survival or indeed for the spiritual awakening and flowering of individual consciousness. That possibility seems to have been lost with the banishment of Adam and Eve from the Garden of Eden and the separation of human beings from the Divine . . . or, at least, from the Creator.

▼   ▼   ▼

There seems to be a disconnect between our physical natures—represented, in this example, by our genetic code—and our consciousness. Consciousness wants to be immortal, to live forever, to maintain its individual identity streams far into the future, far longer than current lifespans would allow. Hence the stories about going to Heaven after death, about individual sins and the punishments for them, about the Last Days and rising from our individual graves and living, as individuals, with God. The genes want survival of the individual for a finite amount of time, to procreate and nurture, and then it is disposed of. Consciousness wants the survival—as long as possible, even after death—of something intangible and difficult to define: our identity. Is there a middle way?

The near-death experience may hold a clue.

You have all heard the stories of those who have died on the operating table and then come back to life after a few minutes. You have heard of the descriptions these survivors have given of a tunnel, and of a light at the end of the tunnel; of seeing loved ones, and of feeling a tremendous sense of peace. Sometimes these survivors return with a feeling that death is not scary or the end of consciousness, but something beautiful and even desirable.

Modern medicine has explained this (tentatively, there is no conclusive evidence) as a natural response of the body's chemical mechanisms, the sudden flow of endorphins designed to facilitate death by making it seem all warm and cozy. The problem with this explanation is that there is no discernible evolutionary reason for such a mechanism. What does the body care how consciousness behaves or what it experiences at the moment of death? The genes have spoken: the body is no longer important, its job is done. What purpose is there in making the death experience pleasurable?

Obviously, from the point of view of the genes alone, there is none.

Unless, of course, death is only the end of Act One in a play in which the principle player is not the genetic code or even matter as we understand it today, but consciousness. The alternative explanation is a little more terrifying: that this "slide into death" made easy by a flood of neurotransmitters is designed by the genes to ensure that we do not seek physical immortality since we are assured of something better after death. After all, individual immortality offers no benefit to the genes and, in fact, may be detrimental to the survival of the entire species as a burgeoning, deathless population consumes all natural resources to the immediate detriment of the younger generations and the ultimate destruction of the species as a whole. Thus, this pleasant near-death experience or NDE may have an evolutionary purpose after all: go back and tell the others that death is not the end, that it is sweet, and that you will be reunited with your loved ones, etc., etc. In this case, the endorphins that flood the nervous system at the time of death are truly the opiate of the people.

In the 1980s a number of studies were done comparing NDE with that of the UFO experiencer.[20] Surprisingly, many similarities were discovered in the way survivors of both types of experience understood life, death, the paranormal, etc. This may seem strange at first glance, but a search through the history of how think tanks and other government contractors approached the problem of the UFO phenomenon reveals that they understood (almost from the outset) that this was a problem essentially related to consciousness.

What may not be well known is the fact that famed UFO researcher Jacques Vallée was himself one of the earliest members of the Stanford Research Institute (SRI) program studying remote viewing (RV), working with Hal Puthoff and Russell Targ and alongside Ingo Swann.[21] He joined the program before the US Army became interested and the whole project was classified, at which point Vallée was no longer directly involved. In his own words he says he was not a great remote viewer—some people have the ability more than others—but that he learned how to do it from Swann and was capable of a rudimentary prowess. We should remember for what it's worth (and the worth may be considerable) that Ingo Swann was a Scientologist (Scientology is considered by historians of religion to be a "UFO religion"),[22] and that Jacques Vallée was once inspired by the philosophy of the Rosicrucians (as was his colleague, Dr. J. Allen Hynek of Project Blue Book fame).[23]

It should be understood that Vallée is one of the foremost proponents of the theory that the UFO phenomenon is an artifact of consciousness itself, a conclusion that he arrived at after decades of research. While he does not discount the tangible, physical evidence of UFOs, he understands that there is much more to the phenomenon than aliens flying around in spaceships. The psychological effects on observers and experiencers is so profound, so life-altering, that this seems to be the whole point of the experience rather than merely a side effect.

And doesn't this parallel the experiences of those who claim to have had visions of a divine being?

Anomalous Mental Phenomena

At some point in the decades-long investigation of the paranormal conducted by the US Army, the US Air Force, the US Navy, the Central Intelligence Agency, the National Security Agency, and the Defense Intelligence Agency using such contractors as SRI and SAIC, among others, the tendency was to find terminology that would satisfy the type of language used in official reports. What the creators of these reports wanted to do was eliminate (as much as possible) what they called the "giggle factor."[24] Terms like ESP, psychic powers, telekinesis, etc., were too prone to abuse and ridicule. Thus, among many other candidates, one of the best characterizations of the whole field became AMP or "Anomalous Mental Phenomena." This would cover virtually the entire range of so-called paranormal abilities, everything from precognition and mental telepathy to psychokinesis or telekinesis: the ability to move objects at a distance using only one's mind. This latter ability was sometimes referred to as "remote perturbation"[25] as distinguished from "remote viewing," and was believed to have numerous military applications. Remote viewing was an intelligence-gathering tool, but remote perturbation would be useful for sabotage, such as throwing switches on enemy equipment, causing it to malfunction, or disabling navigation and control equipment on aircraft, missile systems, submarines, etc. Both were AMP in that only the operator's own mind—consciousness itself—would be employed against the enemy forces.

The science behind these methods was uncertain. Not enough study had been done before the government decided to see their applicability in real-world situations. It is quite possible that the white coat laboratory approach was counter-productive in ways we do not yet understand. While many of these AMP and RV operations were curtailed or shut down entirely after decades of investigation and targeted missions, there is the persistent rumor that elements of the "psychic warriors" either remained on active service or were called back into service, especially after the September 11, 2001 attacks.[26]

▼   ▼   ▼

What does all of this have to do with religion, with consciousness, and with the Phenomenon?

As we said at the beginning of this project, our goal is nothing less than a revolution in the hard sciences as well as the social sciences: a reevaluation of what we know about our function, our purpose in the cosmos, and the potential opportunities and possible threats that exist. Many insist that we are the only life-forms in the universe (certainly, the only ones in our solar system for which we have any evidence at all). It is this very characteristic called "life" that is at the center of the controversy over alien influence and the appearance of unexplained aerial phenomena (UAP). It is also at the heart of the world's religions—monotheist and polytheist—and in order to start discussing what the UFO or UAP phenomena represent we have to begin with defining our terms and locating ourselves in the universe. Until we do, we are taking way too much for granted.

The First Religion

It is generally agreed among anthropologists and historians of religion that the earliest form of what we term "religion" was shamanism. It is a field that largely has been ignored by ancient alien or ancient astronaut theorists who tend to associate alien intervention with the relatively advanced cultures of Sumer and Egypt. Most of what passes for ancient alien theory on the cable channel television shows is concerned with architectural and archaeological "evidence" (the Pyramids, the Nazca lines, etc.), onto which are projected ideas of alien intervention or influence. However, even examples of Neolithic archaeological sites around the world tend to support the idea that there is a cosmological and paranormal aspect to the ancient practice of shamanism that can be interpreted in a manner suggestive of an awareness of extraterrestrial origins.

It is, in fact, with the Neolithic period that we notice a trend away from the type of art found in prehistoric sites that focused primarily on simple geometric figures—straight lines, chevrons, hash patterns—to more elaborate depictions of humans and humanoid-type figures as well as the almost surreal depictions of animals and other creatures. It is also the period when the great megalithic structures began to be erected, structures that have astronomical (scientific) as well as ritual (mystical) functions such as Stonehenge. If one wanted to determine at what point there was some awareness of "off-planet" existence, it could be argued that the Neolithic period presents a clear demarcation between the way in which *Homo sapiens sapiens* understood the world as confined to the earth, and the way in which astronomy and the heavens became more important. It is also the period in which human beings became aware of the unseen aspect of nature, for there is no clear evolutionary purpose to the rituals of the Neolithic shamans unless there was a belief that invisible forces could be approached and manipulated using human activity—work—that did not have an obvious direct relation to a given result. In other words, a hunter has a clear objective: his work is directly related to the killing of an animal and the providing of its protein to his tribe or clan. A shaman's work has no such obvious cause-and-effect characteristics. The function of the shaman, the priest, the sorcerer, the magician, the astrologer, the diviner, is concerned with the unseen. In other words, it is not "rational" or "logical" in any normal sense of those words since it is action without discernible result. It is work, the expenditure of energy, that seems to have no immediate, tangible purpose. Yet, to these hunter-gatherers it was an occupation that was respected. At some point it became understood that the shaman performed a valuable service. At some point it was understood that reality was composed of more than what could be seen or heard or felt; that there were other influences in the world and that certain individuals possessed the capability of understanding and manipulating those influences.

At some point, then, something that ordinarily was *unseen* became *seen*.

However, if we posit alien intervention or alien influence we do not need to restrict ourselves to one period or one culture; contact could have been taking place at various intervals in human history and, indeed, many insist (and the evidence suggests) that this contact remains ongoing. If there was contact during the earliest Egyptian dynasties or during the Sumerian civilization in Mesopotamia, what form did it take? Ancient alien theorists use the religious rituals, iconography and beliefs of these cultures as their starting point and there is a reason for that. As we saw in the quotation from Ufologist Aimé Michel above, religious ideas and alien theories are often related. In the present day, there has been an explosive growth of what have been called "UFO religions." The Phenomenon lends itself to these ideas as it consists of characteristics that are so foreign to everyday experience as to place it in the same category as PSI, NDE, and other paranormal subjects. It is a working hypothesis of this project that what we call mysticism is a necessary adjunct to science when approaching the Phenomenon, as Michel suggested above. This "shotgun wedding" of science and mysticism will not leave either field of study unchanged, however, and we must understand this at the outset.

In the first place, the term "contact" is itself problematic. If we are discussing the many ways in which the Phenomenon manifests itself, actual contact—with the implication of interactions between two species—is mostly what Michel calls "non-contact." The Phenomenon rarely occurs in such a way as to maximize visibility by huge numbers of human beings. There are no landings in major metropolitan areas, in full view of civilians, first responders, and news media. When sightings do take place in cities, they are almost always in the form of unexplained aerial phenomena (UAP) such as in the case of the Washington, DC, "overflights" in July 1952, and not in the form of landings and never in the form of humanoid figures exiting the craft and interacting with humans. Those are Hollywood images. Actual contact between human beings and those associated with UAP have been limited to individual cases taking place in largely remote areas. In other words, in precisely the same environmental context and under the same circumstances as the initiation of the shaman, as we will see below.

What does this imply for the origin of the human experience we call religion but which in this case can just as easily be categorized as mysticism or even shamanism? Did contact take place between human beings and some other lifeform in dim prehistory, or was the experience(s) more in line with UAP?

And is there any substantive difference between the two from the point of view of consciousness? In other words, did the ancient peoples perceive a humanoid presence associated with aerial phenomena in a way that is closed to us moderns? As this Phenomenon partakes of effects that are psychological (we might say parapsychological) in terms that are better described using the vocabulary of religion and mysticism, we are confronted with the possibility that persons within different cultural contexts will perceive it differently, and may see the Phenomenon manifesting in different ways or at least interpret it according to a vocabulary available to them which (ironically) would be misinterpreted by later commentators in light of their own understanding of the terms.

From what we know of shamanism—and particularly those forms for which we have more recent information from anthropologists and travelers in the nineteenth and twentieth centuries—contact with spiritual forces is an integral part of the practice, as is the experience of flight.

This flight may take place to celestial regions, but can just as easily take place to "underworld" regions, i.e., below the earth. The studies seem to indicate that shamanic flight is dangerous under any circumstances; similar to an acid trip which can be good or bad depending on set and setting and the psychological components of the tripper, a shamanic flight seems to be vulnerable to similar forces.

The initiatory experience of the shaman may include such harrowing events as dismemberment and death: experiences that seem real enough at the time but which can be interpreted as analogous to the kind of initiations that took place during the ancient Eleusinian mysteries as well as those depicted in the alchemical illustrations of Renaissance Europe, and during Masonic rituals as well. While initiation implies "beginning," such initiations may take place over a long period and are thereby reinforced each time, albeit in a slightly different context. Freemasonry, as a modern example, provides a First Degree initiation which involves the initiate being blindfolded and initiated at the point of a sword; the Third Degree initiation includes a mock burial and resurrection. There is a progression of initiations that implies a progression of intellectual understanding parallel to the emotional significance of the dramas that are enacted. The dramaturgy is essential to the initiation; the psycho-emotional content of the drama being enacted is necessary to create a psychological space for the symbolic content of the initiation to be "understood," not only in an intellectual fashion (for which there would be no need of a ritual) but also in a way that cuts through conscious awareness of the external circumstances of the ritual to a subconscious realm where the effects of the initiation presumably are felt. The ritual is a carrier of data, a large percentage of which cannot be apprehended intellectually as it involves psychological components that will be experienced differently by different initiates. Compare this to the accounts given by UFO experiencers and the similarities are obvious: there is very little purely intellectual content in the encounters which almost always contain dramatic scenes and emotional trauma.

The temple space used for the Masonic-type rituals is what anthropologist Victor Turner calls a "liminal" space:[27] it is a place set apart from normal, waking consciousness and its very strangeness is its most powerful aspect. The dimensions of space and time are deliberately manipulated to create an otherworldly sense, an emotional reaction to the somber surroundings with incomprehensible symbols and elaborate speeches full of hyperbolic statements and dire warnings. In order to accentuate these feelings the initiate is bound and blindfolded. This creates an element of sensory deprivation and an awareness of one's own physicality as well as the vulnerability of the body to forces outside of one's control (see the works of Michel Foucault on the politics of the body).

From the relatively sedate Masonic rituals we proceed farther afield to the initiations of the shaman. Mircea Eliade has marshalled numerous examples from the available literature on shamanic initiation in his influential *Shamanism: Archaic Techniques of Ecstasy*.[28] While Eliade's analysis has been criticized on various grounds,[29] the references he consulted largely have escaped opprobrium. In these sources, we learn that shamans were individuals who were sometimes selected by other shamans, or who were born into shamanic families. In other cases, the selection process was quite different. Either a potential shaman self-selected, deciding that he or she (there have always been shamans of both genders, depending on the culture, as some Paleolithic graves reveal) wanted to claim that career for whatever personal reason, or the shamans found themselves selected due to other factors such as psychological or physical disabilities which indicated that the potential shaman was already halfway through the "door" to the other world.

When one understands how a shaman functions in society it becomes clear why illness should be a positive factor in selection. Illness is the poor man's altered state of consciousness. If it is a physical illness, it demonstrates to the sufferer that the body is unreliable and subject to changes, thus placing the sufferer in a different social category and conditioning him/her to think differently about the world and to suspect that there is a hidden or occult aspect to reality. Anyone who has been a patient in a hospital, for instance, realizes that they are in a world that is separate and apart from normal, everyday reality. Schedules are different, food is different, there is a different cast of players to deal with, all imbued with the realization that death is a distinct possibility. It is a world we never consider as part of our lives. Even as visitors to a patient in a hospital, we cannot appreciate the separation and isolation from the everyday world with all of its pleasures, pains, and conceits that takes place once you are in a hospital bed and no longer have control over your own life but instead have been rendered helpless with all of your bodily functions a subject of scrutiny by strangers.

If it is a psychological disorder, then of course we are on track to understanding that reality is not what it seems and that other modes of living and being in the world are possible, if unappreciated by the world at large. The psychiatrist R. D. Laing became famous—some would say notorious—in the 1960s for suggesting that a nervous break*down* was also a nervous break*through*.[30] His characterization of schizophrenia as tantamount to spiritual initiation went through a vogue briefly before it was superseded by a psychiatry that was more interested in chemically controlling the mind of the schizophrenic than in trying to communicate with the patients through analyzing their symbolic language. It was thought that the chemical approach was as close as anyone was going to come to a "cure" of mental disorders and that since behavior could be changed using drugs it was decided that certain mental disorders were nothing more than chemical imbalances in the body: a case of making the nail fit the hammer. This was Laing's "politics of experience": the effort by medical institutions to come up with a socially acceptable means of identifying, characterizing, and treating psychic states as disorders rather than as spiritual quests, which has led some observers to consider the shaman as a psychopath.

A shaman is expected to travel to the spirit world on a regular basis; a shaman communes with spiritual forces and is in some cultures possessed or inhabited by a spirit who is revealed when the shaman is in a trance. In some cultures, this trance is autohypnotic in nature: the beating of the shaman's drum or the sound of the rattle may be enough to send the shaman into a hypnagogic state. In other cultures, entheogens—hallucinogenic drugs in the form of plants or fungi—are used to bring the shaman into contact with the "other world."

This other world is a field of symbols: not only the static symbols of icons and images, but active symbols that manifest through sound, dance, the play of environmental forces, and the hereditary narrative of the village itself. The shaman is the living repository of the history of the clan or tribe; as such, he or she can communicate in a kind of sacred language, a shorthand of symbols and semiotic content that is comprehensible to the other members of the tribe and can be interpreted by them while the shaman is still in a trance.

We see this in the contemporary example of the state oracle of Tibet.

This is a tradition that has very little to do with Buddhism proper. It is a legacy of the indigenous religion of the area which is generally known as Bön. The origins of Bön are unknown, but tradition has it that it came to Tibet from a mysterious land far to the west known as Zhang Zhung. However, the position of the state oracle is an important one and the Dalai Lama has been known to consult with the oracle in the context of a very elaborate ritual during which the oracle becomes possessed by a spirit and utters predictions in a coded language. It is safe to say that this represents an instance of shamanism. The person who is possessed by the oracle is the head of the Nechung Monastery in Tibet, now in exile with the Dalai Lama in Dharamsala, India. He is known as the *kuten* or "material basis" for the spirit oracle but only when in the context of the specific ritual that is used to summon it. In that ceremony, the monk is dressed in an elaborate costume that weighs in excess of seventy pounds; under normal circumstances the man hardly would be able to walk in it. However, during the ritual, the possessed monk becomes quite active, moving freely around the temple to the accompaniment of chants and music played on a variety of instruments.

It is interesting to note that the monk does not perform as a shaman except during the ritual used to invoke the oracle. Ordinarily, while he is certainly a spiritual leader, he does not act in the role of shaman but only as the head of a monastic order. However, his function as oracle does take its toll and he has been known to require convalescence after each possession, in some cases finding himself unable to walk for long periods of time.

This gives us a laboratory case for understanding one aspect of the Phenomenon: the permeability of the veil that separates this world—the world of everyday, waking reality—from another world. We may call this other world a parallel universe, we may call it the aether, the astral plane, or whatever form is most congenial. Until we have developed our science to the point where we can know more about it, almost any designation will do. The problem is that many of these terms are so culturally loaded that they wind up carrying much more information and implication than they can handle. We would need to deconstruct or at least unpack these terms before we would be comfortable in using them.

The term astral plane is quite suggestive, and appropriate in some ways to this project if we can extricate it from some of its more New Age associations. The word "astral" means, of course, "starry." The astral plane is the plane of the stars. The concept is ancient, and goes back at least as far as Plato and then was expanded by Plotinus and the Neo-Platonists. It goes back to the idea that the Earth was the center of the universe and that around it were the orbits of the planets and the stars. One ascended to the heavens through the plane of the stars. Normally one did not do this in the body; the body stayed on Earth while the soul moved through the planes and then came back with information or insight. This is the "journey" of the shaman, codified and made systematic by the ancient Greeks.

While a human soul may travel outside the body to walk among the stars, the reverse was also true: spiritual beings could appear on the Earth if they were able to possess a material basis. The Tibetan oracle is one such material basis: a vehicle for the spirit to inhabit for a short period of time. Today we might call this "mediumship" or even "channeling," in a kind of secularized version minus the complicated rituals. All this goes to show that there has existed in human culture for millennia the concept that there are at least two worlds—this one, and one other—and that they co-exist; travel between the two is possible, as is communication between the two. This is the privilege of persons who have a natural ability for this kind of commerce, or for those who have been trained in its processes.

The medium or the shaman makes themselves available to spiritual forces who wish to penetrate into this world from their own; but what of spiritual forces who do not require the cooperation of human agency? What of those spiritual forces who just . . . show up?

One of the hallmarks of the Phenomenon is that there is rarely an effort by human beings to "call" them to visibility. The aerial phenomena just happen. They are unpredictable and unrepeatable, and thus they resist scientific testing which, by extension, means that they do not exist. It's a vicious circle, a function of the scientific method which demands repeatability and predictability of scientific phenomena.

There is, however, a precedent for this although it has not been addressed before since the entire subject resists any kind of sober academic study aside from sociological treatises on "UFO religions" and the like. The precedent is the European practice of ceremonial magic.

▼   ▼   ▼

A close relation of shamanism in some ways, ceremonial magic is predicated on the idea that a human agent can penetrate the spiritual realms while conscious and at a time and place decided in advance. Ceremonial magic uses many of the same basic principles as the ritual of the Tibetan state oracle described above, with the major difference being the magician does not go into a trance or lose consciousness. The magician does not become a "material basis" for the spirit; instead, the magician penetrates the veil that separates the proposed two worlds to summon a spiritual force to *visible appearance.*

The way this is done is suggestive of some sort of familiarity with aerial phenomena. The typical arrangement as attested in the grimoires—the manuals of the magicians— is to draw two concentric circles on the ground. Between them is written a variety of magic words, usually in Latin, Greek or Hebrew, as well as hieroglyphs that may be the signatures of various spirits or the seals of spiritual forces that will protect the magician and ensure obedience from the spirits being summoned (or "conjured"). The magician stands in the center of the circle, and there is usually a minimum of four candles or lamps, one at each of the cardinal points, as well as a censer of burning incense.

Prior to the ritual proper, the magician is prepared with fasting, prayer, and other measures to focus the mind and create a state of sensory manipulation in which the only sensory input comes from the carefully constructed ritual. More than just a liminal space, the ritual area reflects the precise nature of the magician's goal. The colors of the candles as well as of the furnishings of the room or environment where the circle is prepared, scents (the incense), sounds (the incantations, the ringing of a bell in a precise number, etc.), have been arranged to reflect the purpose of the ritual.

The magician is dressed in robes appropriate to the ritual, as well. One is familiar with the strange costumes of Merlin, or of Disney's *Sorcerer's Apprentice*. There may be an exotic headdress, symbols embroidered on the robe, a wand, a sword, an amulet or talisman, etc. In this way, the European magician shares some qualities with the Siberian or Tibetan shaman for whom the costume is an essential part of the ritual, further stressing the liminality of the ritual, but which also may be an allusion to the forms of another—more alien—environment.

The ritual begins with invocations to God, the angels, etc., and proceeds to conjurations of the spiritual forces. This may be accompanied by a specific series of knocks, or knells (strikes on a bell or gong, the number of which corresponds in some way to the nature of the spirit being summoned), chanting of *voces magicae* or what is known as "abracadabra": seemingly meaningless syllables that have the effect of transcending normal language by robbing sounds of meaning (or attaching arcane meanings, symbolic meanings, to sounds). Ritual gestures may include the sign of the Cross, or of the Pentagram, etc. In other words, the entire field of action and sense is managed and tailored to the goal of the ritual. There is no wasted movement, no extraneous sounds or sights or smells. Thus every event that takes place under these circumstances that *has not* been determined in advance—every "random" occurrence—is evidence of some form of contact.

The proximate goal of this ritual is to force a spiritual entity into visible appearance. The magic circle serves to protect the magician from hostile spiritual forces and also reinforces the magician's position at the center—the axis mundi—of creation. In the ritual, the spirit—angel, demon, planetary intelligence, etc.—appears outside of the circle, sometimes within a specially designed and drawn triangle at some distance from the outer rim of the circle. This triangle may also contain lamps, burning incense, etc., and the spirit constrained to appear within it. The spirit then answers questions put to it and eventually is given the license to depart. At that point the ritual is over.

If we apply the Cargo Cult metaphor to this ritual a few things become apparent rather quickly. The concentric circles with the hieroglyphs and strange lettering between them can easily stand in for the aerial vehicle, a "flying saucer" if you will, especially when we factor in the lights and the smoke from the incense, the odd sounds, and the oddly clad passenger within. Even the triangle is not unknown to UFO lore as the contemporary sightings of triangular-shaped craft in Belgium have suggested.

Strange signs and symbols are a part of many UFO sightings, perhaps the most famous being Roswell's debris which contained a series of such symbols (according to some witnesses), but it is not the only case. Files on the

Phenomenon contain a plethora of references to hiero-glyphics or strange lettering—sometimes misidentified as Cyrillic or Sanskrit or Egyptian—and when occupants have been seen they virtually always are dressed in strange clothing and are said to communicate telepathically.

One of the oldest of the European grimoires is known as *Liber Iuratus Honorii*, or *The Sworn Book of Honorious*. This is an early fourteenth-century text that contains rit-uals for summoning spirits in the manner that has since become familiar as ceremonial magic. Whereas other magic texts—such as the eleventh century *Picatrix* and various forms of the *The Key of Solomon*—predate the *Sworn Book*,[31] the latter gives us what would become the template for a standardized rite involving magic circles, long ceremonies, complex preparations, etc.

It is well worth noting that this earliest of grimoires stipulates that the magic circle be created out of bricks or stones, with a mound of earth in its center.[32] This is the form used for summoning the *planetary* spirits. The spirits would appear on the mound in the center of the circle of stones. In other words, the magic circle with the mound in the center is the vehicle for the *spirit*, not for the magician (who draws another set of circles for himself).

And its shape is virtually identical to that of the iconic "flying saucer."

▼   ▼   ▼

Now perform a kind of thought experiment in which you reverse this ritual. It is no longer a human being standing in the center of the magic circle, summoning a spirit. It is an "alien" standing inside what we think of as a flying saucer. This alien in its saucer penetrates our world—our "astral plane," if you will—to summon *us*. The appearance of the saucer is a manifestation of someone in a parallel universe using ceremonial magic to enter our world.

This is an outrageous suggestion, to be sure. Yet, consider this: In many accounts of close encounters—some of the "high strangeness" that accompanies this phenomenon—unsuspecting individuals hear knocks, or strange sounds, see eerie lights, and feel that they are somehow in telepathic communication with something not quite human.

The instructions given in the grimoires usually emphasize that it is necessary to command the spirits, to maintain the upper hand at all times. The magician is normally provided with weapons to ensure obedience and these are magic swords, knives, even scimitars. The spirits are threatened with these devices if they do not respond or if they appear to become hostile or disrespectful. Water that has been blessed by a priest can also be used for the same purpose, or specially created sigils drawn on virgin parchment with the names of God and his angels. In other words, there is an element of confrontation and the threat of violence in these encounters.

In many cases where there have been close encounters, human beings have felt threatened by the visitors; they have experienced pain, anxiety, claim to have been subjected to physical abuse of various types, even to have born burns or scars as a result of the encounters. Even stranger, some experiencers claim to have been abducted by the alien entities on multiple occasions. This could be interpreted as having been summoned by a magician and forced to do his or her will. In the old texts, it has been suggested that the magician force the recalcitrant spirit to obey through force and the use of pain—the point of a sword, for instance—and by extorting an agreement to "come again when called."

The magician finds that a strange space has been entered; the magic circle is a vehicle for entering that space as well as a means of protecting the magician. If we think of the magic circle as a craft for traveling from this universe to a parallel one, we can see that the "control system" would be unknowable to a lifeform on the other side. The manner of propulsion would be a mystery; and in order to move from our universe to the one next door we would necessarily violate all known laws of physics. The travel would be conducted in a kind of trance, a state of suspended animation in this universe but a dynamic motion through the next (albeit undertaken with a large amount of ignorance as to the environment of that universe). And should the ritual be successful, beings in the other universe would notice our magic circle, perhaps consider it a kind of vehicle, and would be terrified or fascinated by it, or both.

The reason for this thought experiment is to suggest ways in which we might understand the Phenomenon from a completely different perspective. There is no implication or insistence that this experiment is in any way "true," but it does serve to open the door to alternate theories of the Phenomenon by inserting more specifically paranormal and mystical ideas into the discussion.

Jacques Vallée has suggested that the Phenomenon represents a kind of control mechanism, and this is precisely what the ritual just described intends to be. The magician must be in complete control of the ritual and by extension of the beings he or she summons to visible appearance. This idea of "visible appearance" indicates that the membrane separating our two universes can be penetrated or breached, if only for brief periods of time. During this ritual both the magician and the spirit evoked are visible to each other; two worlds have come into contact and bled into each other, if only slightly, but enough to cause an important change to take place in the mind of the magician if not also of the spirit.

In other words, *we* may be the "spirits" for the magicians who fly the ships.

▼   ▼   ▼

Using the vocabulary and analogy of ceremonial magic may seem like a stretch in our attempt to understand the Phenomenon but it should be noted that the American military and intelligence organs did not hesitate to probe every type of occultism, parapsychology, mediumship, etc., in the quest to weaponize the psyche and to understand the Phenomenon. Basically there were two separate streams of research that had their origins in the 1950s with the onset of the Cold War as well as the war in Korea with its associated fears of "brainwashing." The first stream is represented by the military and its growing preoccupation with the potential of psychic abilities to probe enemy secrets as well as sabotage enemy installations. The second stream was that of the Central Intelligence Agency and its mission to uncover the secrets of the human mind for use in psychological warfare, behavior modification, and mind control or "brainwashing." We will go into this area in more detail in Book Two. It is sufficient in this place to acknowledge that what we are attempting to do here is suggest a spiritualized refinement of those methods in order to maximize their benefit for the common good in terms of our proposed revolution in the sciences and culture. The term "spiritualized refinement" should not be misinterpreted: we are not applying any kind of doctrinal or dogmatic ideology to this process. We are in rough agreement with those whose distaste for religion comes from the abuses committed by organized religion throughout world history. Churches, after all, are human institutions like governments and are as susceptible to corruption, abuse of power, and all the other negative effects of placing temporal authority in the hands of other humans. Instead we propose—in a manner reflective of the best minds currently working on this problem—what may be termed a (in lieu of a better word) "mystical" approach that is devoid of any denominational influence; moreover, a mystical approach that is seen as a companion and partner to the purely materialistic approach of science.

This is the only way in which we can begin to understand the true nature of the Phenomenon since it reveals itself to us in both of these dimensions.

The Demiurge and the Alien God. The material world and the world of spirit. Science and mysticism. Rational and non-rational. The earthly and the astral.

The human and the alien.

A serpent with a lion's face, a solar disk, and symbols of the moon and stars. This is one possible rendition of the Gnostic idea of the Demiurge: the "real" creator of the world. The similarity of this symbol – a serpent rising between the moon and a star – would be replicated in esoteric imagery around the world, including in the famous system of Kundalini yoga.

The Death of Simon Magus, from the Nuremberg Chronicle of 1493. Simon Magus was a Gnostic magician and according to some texts an early opponent of Saint Peter. In this image, Simon Magus is being cast down from the skies over Rome where he has been engaged in a magical battle with Saint Peter to see whose spiritual powers are stronger. Simon, whose assistants are demons, is being attacked by an angel summoned by Peter. This celestial conflict between two classes of beings and their human surrogates is a constant theme in magic and esoterica.

This is a painting by William Blake of the famous scene in the First Book of Samuel (28:3-25) involving the Witch of Endor – a magic ritual specialist – and King Saul who demands that the Witch summon the ghost of the prophet Samuel. Compare to the following:

Joseph Smith, Jr., the founder of the Church of Jesus Christ of Latter-Day Saints (Mormonism) evoking the Angel Moroni on the Hill Cumorah in upstate New York, the end result of a series of magical rituals that took place over several years.

Compare to:

EDWD KELLY, A MAGICIAN.
in the Act of invoking the Spirit of a Deceased Person.

The conjuration of a spirit of a deceased person by Edward Kelly and John Dee, two Elizabethan magicians. Dee and Kelly were on a prolonged series of magical operations designed to speak to angelic forces, in the course of which they were presented with an "Angelic Language", sometimes called "Enochian", with its own alphabet, vocabulary and grammar. This language is used to this day by magicians and occultists.

This is a drawing of a shaman from Lapland with a magical drum. The drum is covered in occult symbols intelligible only to the shaman. A shaman undergoes a rigorous period of psychological disorientation and physical isolation in order to prepare for spiritual initiation, after which the shaman will be able to travel to the stars to speak with the spirits. Compare this concept with that of Simon Magus, depicted earlier.

A photograph of a shaman from British Columbia, Canada in the early twentieth century.

A photograph by Christopher Michel, showing the State Oracle of Tibet in a trance, on his way to advise the Dalai Lama.

A magic circle from the *Heptameron* of Peter of Abano. Notice the concentric circles with some arcane symbols between them, and the four stars at the four corners. If one were to view this in three dimensions instead of two, the similarity to the design of the archetypal "flying saucer" would become apparent. Indeed, the magic circle is a mechanism for visiting other worlds.

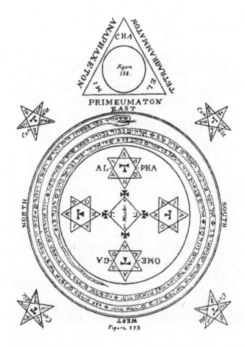

A variation on the magic circle, this one from the fabled *Key of Solomon*. Notice the presence of a coiled serpent, four stars, and now a triangle to the east of the circle. Each of the four stars would hold a candle. The orientation of the circle, stars and triangle is of prime importance, suggesting a connection between this world and the celestial.

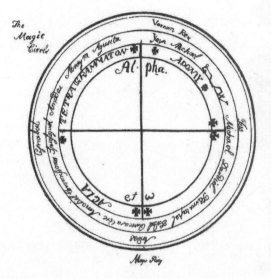

This is a magic circle from the book by Francis Barrett, *The Magus*, published in 1801. A compendium of occult lore from diverse sources, such as Cornelius Agrippa, it was the volume most likely consulted by a young Joseph Smith in his conjurations in the woods outside Palmyra, New York.

THE GATE OF THE GODS

From time immemorial, man has felt himself to be confronted with evil supernatural beings . . . Amid such fears and wonders lived the river peoples of the Tigris and Euphrates: the legendary Sumerians . . . On broad plains, on terraces of temples and towers, the priests scanned the night sky, pondering over the riddle of the universe–the cause of all being, of life and death. . . . By conjuration, by the burning of incense, by shouts and by whispers, by gesture and by song, the priests sought to attract the attention of the fickle gods . . . "Spirit of the Earth, remember! Spirit of the Sky, remember!"[1]

THAT THE EXPERIENCE WE ARE CALLING THE "Phenomenon" was known to ancient civilizations just as it is to our own is as much a fact of history as Kenneth Arnold's sighting in 1947 or the thousands of similar sightings that have taken place all over the world since then. To demonstrate this—and to get some idea of the earliest direct experience of the Phenomenon in recorded history—it is necessary to go back to what is arguably the world's oldest known civilization, whose writing has come down to us and has been translated: the Sumerians.

The Sumerians will give us some context for an understanding of how human beings have interacted with the Phenomenon when there were no internal or external censors to alter the reports or confuse the issue with political agendas. That does not mean their accounts are reliable from the point of view of scientific inquiry; what they experienced was as difficult to describe for them as it remains for us. Yet, the consistency of their emotional and cultural response to the Phenomenon—or to the memory of the Phenomenon which was handed down from earlier generations—gives us a stable platform from which to begin our investigation.

Further, we will see that there is enough similarity between the ancient Sumerian and Babylonian records on the one hand and the biblical record on the other that we can use the one to help understand the other. The pre-biblical accounts are a polytheistic version and the biblical account is monotheistic, but they share many essentials aside from their ideological differences. The importance of the Sumerian originals is that their legends give us the oldest written evidence of what we may call "contact."

The Two Worlds Interpretation

One of the essential characteristics of the Phenomenon is its "otherworldly" aspect. The Phenomenon does not seem to originate in our world. This raises a question: what *is* our world? How do we determine what is part of it, and what isn't?

The answer may seem obvious: whatever comes from the Earth is part of our world. If something does not come from the Earth it must come from elsewhere, from another world. However, things are never that easy.

The concept of "world" has changed considerably over the millennia since earliest recorded history. The world was whatever was known and discovered in the course of the rise and fall of civilizations. Alexander the Great was considered to have conquered most of the known world of his time, stopping only at the border of India, even though there were vast territories—including the entire western hemisphere—that he did not know existed. The Roman Empire stretched from Palestine and Egypt to the British Isles, but that was nowhere near the whole planet. Old maps would show blank areas outside of whatever empire in which they were designed. The western hemisphere was not part of the world until it had been officially discovered in the fifteenth century CE, even though there had been expeditions to the eastern seaboard by the Norsemen before that, and even though the Clovis people (and earlier migrations) had traveled across the Bering Strait land bridge and down into North and South America long before there was an Egyptian dynasty or a ziggurat at Ur. Today, we believe we know the limits and boundaries of our world but that confidence depends on an old paradigm that is largely geographic and spatial and which does not take into account other dimensions of experience and of theories of modern physics that challenge our comfortable sense that we know our world and our position in it.

In fact, the English word *world* is itself somewhat problematic. It has a Germanic origin and comes from two words: "man" and "age." Thus, a "world" is an "age of man."

This gives the concept of "world" a decidedly anthropocentric spin as well as focusing on the idea of age, and thus of time. In Sanskrit, an "age" is a *yuga*, similar to the Gnostic concept of *aeon*. The age we are living in now is called the *kali yuga*, the "age of Kali": a time that is also a personification and a deification.

For the Romance languages, the words *mondo*, *mundo*, *monde*, etc., derive from the Latin *mundus* which means "clean, neat, elegant" and "refined" as well as "ornament." This is said to derive from the Greek *cosmos* which means "orderly." In other words, this understanding of "world" indicates order, neatness, even elegance. A world is something that has been designed perfectly, with associations of symmetry and even beauty.

What, then, is the *other* world? If we are to go by the assumptions in the English and Romance language interpretations the other world would be the domain of the non-human, the dirty, the messy. The age of man would become the age of non-man. It would be a time as well as a space, populated by beings of which we have no knowledge or recognition. Something outside the symmetry of the beautiful world.

Thus, the "world" is not the domain of everything the way we usually think it is. We may visualize a Venn diagram in which "world" is one circle, and "other world" is another circle. When it comes to the Phenomenon, the two circles overlap in a small area: that region is the domain of the Phenomenon, called in Mesopotamia the Gate of the Gods, or Babylon.

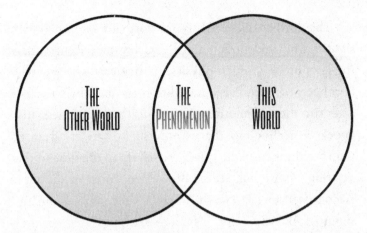

Now this idea is predicated on the assumption that there are only two worlds, this one and another. That is by no means certain; there may be multiple worlds, not only multiple dimensions or multiple universes, but domains in which the physical laws that seem to apply in our world seem similar to ours but modified or altered in some way.[2] If this "other world" were entirely different—a parallel universe, for instance—it might remain invisible (or otherwise undetectable) to ours forever even as it impinges on our own. What we suggest is a world that is enough like ours that it provides some degree of reference. That is why the Phenomenon appears to us to be almost "real," almost contemporary . . . but not quite.

What is another characteristic of the Phenomenon? That it is visible. The visual sense is the first of the five classic senses—according to biologists and paleontologists—that appears during the process of the evolution of consciousness. The auditory, olfactory and other senses involved in consciousness arrive somewhat later.

The third characteristic is one normally associated with sightings of unidentified aerial vehicles: they seem to defy the laws of Newtonian physics, particularly where gravity is concerned. However, in addition to that obvious characteristic there is another that should be mentioned: they appear as technology that is only slightly more advanced than anything humanity had devised up to the time of that sighting. As we will see in Book Two, some of the more famous sightings in North America—the famous "airship" sightings of 1897, for instance—demonstrated technology (or technological design) that was perhaps twenty or thirty years (at most) beyond the capabilities of 1897 manned flight. There is thus a time factor involved in the Phenomenon that is difficult to understand, but it exists.

The visual information that is gathered during these sightings is the beginning of a kind of communication; it indicates an interface between the two worlds (as noted above), one that will require a sophisticated definition of consciousness because it is the meeting place between two conscious beings that are otherwise dissimilar. Whatever is causing the Phenomenon to exhibit these characteristics has "stepped down" its visual impact so that it teases at the edge of our imagination.

"Imagination" is the key concept here, for whatever caused the 1897 airship sightings did so by making an appearance that was *almost* credible to the level of technology of the general population at the time. They looked like flying boats, with various contraptions and appendages hanging off of them whose function was mysterious.

They were often silent, but could change direction suddenly without an obvious propulsion system. They were just advanced enough to cause wonder and astonishment (if they were gas-filled balloons no one would have noticed, and it seems important that they be noticed), but not too much so that they would be virtually invisible or beyond description. In other words, the Phenomenon is not looking to be ineffable, just wonderful. Fantastic, as in a manifestation of fantasy. It is as if we are being urged to dream a little more daringly, to imagine the possibilities of which we are capable.

They have baited a hook, which we imagine to be a worm, and we take the bait.

▼   ▼   ▼

In Sumer, a similar situation occurred.

By reading through their creation legends and some of the other "otherworldly" texts produced by the Sumerians, the Akkadians, and the Babylonians, and by examining their monuments and other architecture, we can see that they were similarly affected by things seen in the sky. Their gods and cultural heroes went on journeys to the Underworld and to the Heavens; their high priests met with incarnations of a goddess on the top of their ziggurats at certain times of the year (anticipating the contact between the high priest of the Israelites and their God once every year, on Yom Kippur, in the Holy of Holies in the Temple of Solomon). As stated above, the earliest civilization in the world with any kind of written documentation—the Sumerians—provide the template for much that will follow, and will introduce us to the concept of two worlds and the phenomena that accompany them. And they will lead us to ask an important question, one that may be at the heart of government secrecy concerning the Phenomenon: Who's in charge?

▼   ▼   ▼

By the power invested in . . . whom?

Oddly, as long as the Earth was considered the center of creation, the authority held by human rulers was said to derive from the gods. Once the sun was moved to the center (at least, of our planetary system), then authority shifted from the gods to human beings and rulership became anthropocentric. This seems somewhat counterintuitive. Just when we realized that we were not at the center of the universe, our science, philosophy, and culture began to insist that humans were more important than the gods. Religion became a quaint holdout against this anthropocentric point of view, even as human beings themselves seemed less and less capable of running anything for any length of time without screwing it up.

The Sumerians were deeply concerned with the issue of divine authority. The Sumerian rulers were believed to be in contact with the gods through the enactment of rituals which reinforced their authority in the eyes of their people even as they "met" the gods on specific days of the year.

Each Sumerian city had its ruler, who was both secular leader as well as high priest of the Sumerian mysteries. Further, each Sumerian city was ruled by a different god.

The center of religious worship was the ziggurat: a stepped pyramid, usually surrounded by a brick enclosure. There were chapels devoted to lesser gods, but at the very top of the ziggurat was a small chamber or temple of glazed brick. This was the holy of holies, where the god of the city would descend and meet with the the city's ruler. In the case of Babylon, the ziggurat had seven levels and the top chamber was of blue glazed brick where Marduk would appear one day a year to a high priestess designated for a ritual marriage.[3] If Babylon was the "Gate of the Gods" then this chamber of blue glazed brick at the top of the ziggurat at the center of the city was the point of tangence between the two worlds, the one small space on the planet where god and human met face to face.

Various New Age authors have emphasized the importance of Sumer and its rituals, scriptures and architecture to the "ancient alien" hypothesis, and there is an element of truth to these assertions, although it may not be obvious at once. While Zecharia Sitchin has written extensively about his beliefs that the Sumerian concept of the Annunaki refer to alien technicians, most contemporary scholars of Sumer find no substance to these ideas, citing poor translations of Sumerian texts as well as misinterpretations of Sumerian scriptures and conflation between Sumerian and later Babylonian texts. Sometimes, though, the most revealing evidence for non-human or non-terrestrial contact is hidden in plain sight. In this case, it can be found in the Babylonian creation epic, the *Enuma Elish*, and the story of Marduk himself.

Most are familiar with the Creation story as it appears in the Book of Genesis. In that version, God creates the heavens and the earth and eventually creates Adam, the first man, and Eve, the first woman. This story is told in different ways in the Bible—even within Genesis itself—and it is believed that this seeming inconsistency in the accounts is the result of two separate versions (the Priestly version and the Yahwist version) being included. That the Creation story in Genesis is a reframing of the Mesopotamian original with a view towards presenting a monotheist interpretation as opposed to the prevailing polytheist version we find in most Creation stories in the Middle East is the subject of some debate among biblical scholars today.

The biblical process of creating the universe and then the humans who inhabit the Earth is a relatively benign one: there is no violence involved in Genesis 1, no struggle between opposing forces. In fact, in Genesis, God appears as a single uncreated entity with no consort or female companion;[4] yet he is anthropomorphic enough that he can walk through the Garden of Eden and speak with Adam. This is explained by saying that God created Adam in his own "image and likeness." Thus, human beings are somehow reflections or versions of the divine Original. The character or extent of this likeness is nowhere speci-fied, however. Do humans look like God? Do they possess faculties similar or equivalent to those God possesses? We only are told that we share something in common with our Creator. Indeed, when God creates Adam from dust, he breathes life into him. This divine spirit may be the key to understanding how human beings are made in the "image and likeness" of God. It may be a way of explaining consciousness.

The scholar Alan F. Segal has written[5] about the fact that for a long time there was a tradition among the Jews of "two powers in heaven": there was the invisible, transcendental God that we know from the Bible, but also a kind of vice-regent of God, a visible, largely anthropomorphic entity that could walk and talk among the created humans. It was only much later that this belief was "corrected" and the idea of a vice-regent officially excised from Jewish theology as a heresy. In fact, it was still a common understanding during the time of Jesus, which is one reason why pious Jews could contemplate the idea that God could look and live like a human being and dwell among them. In this case, it is clear that "image and likeness" means just that: at the very least, human beings are simulacrums of the divine original.

In the Babylonian Creation story, the *Enuma Elish*, there is a long, protracted battle between various gods that finally results in the creation of human beings. This story pre-existed the biblical version by hundreds, if not a thousand years. It is now generally believed that the Pentateuch—the first five books of the Bible, also known as the Torah or the Books of Moses—was written or compiled in the sixth century BCE. The *Enuma Elish* has been dated to anywhere from 1000 BCE to 1800 BCE at the earliest.

In the *Enuma Elish*, divine powers existed in a primordial state, before the universe was created. Foremost among these was Apsu and his spouse, Tiamat. Tiamat is described as a serpent or dragon deity, connected with the waters that pre-existed the land and the stars. Tiamat represented salt water, and her husband Apsu (or Abzu) represented sweet or fresh water. It was the comingling of these waters that gave rise to the generation of gods. As pointed out in several sources,[6] Apsu and Tiamat are not gods per se, but seem to represent proto-deities: perhaps natural forces out of which the Sumerian pantheon was born. We know from Genesis 1:2 that God is depicted as hovering or moving over the "face of the waters." Thus there is some general agreement between the two texts that water pre-existed the rest of Creation. This sentiment is echoed in the Hebrew word for "the deep" or "the abyss" which is *tehom*, a word cognate with Tiamat.

The similarity ends there, however. The Middle Eastern scriptural texts usually incorporate the motif of a *chaoskampf*, or struggle against chaos, that gives rise to order and the created universe. This struggle takes place between gods, with one "set" of deities winning in battle with another "set." The *Enuma Elish* epic is no different, and may, indeed, represent the earliest recorded version of this type of narrative. But because the Bible is a monotheist text, there can be no chaoskampf where Creation is concerned, for that would indicate the presence of multiple deities and an ensuing struggle between them that would imply a certain degree of equality. In a monotheist context, there can only be one god and that god must be all-powerful: thus, there is no space for any kind of struggle. We may say that the biblical version of Creation is a deliberate ideological challenge to the prevailing, polytheist view and does not represent the method of Creation (particularly the creation of human beings) as it was understood by most cultures of the ancient Middle East. However, there was no escaping several basic elements that are common to polytheism (and which hint at non-human actors) which have given scholars headaches, as we will see.

In the *Enuma Elish*, the younger gods—the children of Apsu and Tiamat—are noisy. They are creating enough disturbance that Apsu wants to destroy them. Apsu, however, is killed during his confrontation with the younger gods and this angers Tiamat to the extent that she creates eleven hideous monsters to do battle with the other gods. The leader of the resistance against Tiamat is Marduk.

Marduk is the god worshipped specifically at Babylon, dating from the period when Babylon was not as powerful as some of the other Sumerian cities to the time when Babylon controlled much of Mesopotamia making Marduk even more important. There is evidence for the existence of the worship of Marduk (called AMAR.UTU in Sumerian, or "Calf of the Sun") as early as the third millennium BCE, which is roughly the period of the earliest examples of Sumerian writing. The Sumerian people most likely settled in Mesopotamia around 4000 BCE or earlier, and it is considered likely that they migrated there from what is now Bahrain. It is not certain whether AMAR.UTU was a god they brought with them from Bahrain or if he was a Sumerian version of a local Mesopotamian deity whose culture is now lost. He does, however, have associations with water, with astronomy and the calendar, and with magic.

Marduk combined in himself elements we would find distributed among the gods of Egypt, for instance. He was part Thoth (Tahuti), part Horus, part Amun-Ra; and after his defeat of Tiamat in the great cosmic battle that gave birth to the cosmos, he became the ruler of the divine pantheon. He is sometimes said to have been born in the *apsu*, the abyss, indicating that he is a direct descendant of Apsu, the partner of Tiamat, and not actually born of the union of Apsu and Tiamat. He sometimes was considered the son of Enki, another deity associated with water or an underground ocean, whose temple was known as *E-apsu*, or "house of the abyss." Without delving too deeply into the Sumerian epic it is enough to mention that Marduk became the hero of Creation by defeating Tiamat, capturing her eleven monsters and her consort Kingu (who replaced her first husband, the slain Apsu) and slicing her body into halves: one for the heavens and one for the earth. He then sets the stars in their courses and establishes order out of the chaos.

What is often neglected in the popular telling of the *Enuma Elish* is the story of the Tablet of Destinies. This is the device that gives Marduk his authority over the other gods, the possession of which indicates his kingship. It is the mark of sovereignty, and according to the *Enuma Elish* is worn as a breastplate. Its name—Tablet of Destinies—implies a divination or fortune-telling function as well. Marduk feels entitled to the Tablet because Tiamat has given it to Kingu, whom she married upon the death of Apsu. Kingu is a weaker and somewhat effete replacement for Apsu, and does not seem to merit the authority that the Tablet confers or recognizes. Marduk, through sheer force of character—perhaps what sociologist Max Weber called "charisma" after the Greek word for "gift" or "grace," an inherent quality that is not earned but which is intrinsic, and which may be related to the Sumerian concept of *me*—is depicted as being the rightful heir of the Tablet of Destinies.

It is this idea of sovereignty and authority as deriving not from humans or human institutions but from otherworldly sources that suggests trace evidence of contact. In this case, authority over human institutions is not conferred by humans themselves. The authority derives from elsewhere. Even Marduk, a god, does not automatically receive the Tablet of Destinies by right of birth or origin but must win it by killing the holder of the Tablet, Kingu. A close reading of the *Enuma Elish* suggests that Tiamat committed a grave error by giving the Tablet to the grossly unworthy Kingu, a situation that contributed to her own defeat.

There is another tradition concerning the Tablet of Destinies: that it was the sum total of human knowledge in written form, designed to withstand a deluge. Written in stone so it would survive almost any environmental disaster of the time, it represented the continuity of information from the antediluvian to the post-diluvian period. Thus it was wisdom, knowledge, and authority: a single device, without which human civilization would not be able to jump-start itself after the Flood.

This idea of a written document that has been buried in order to survive a catastrophe is a very old one and found in many different cultures worldwide. There is the tradition of the Emerald Tablet of Hermes: a stone carved with the secret of the universe, buried in a cave in Egypt or under the sea, found centuries later and resurrected from its hiding place. There is also the Tibetan tradition of the *terma*: esoteric texts that were buried by teachers in ancient times and then rediscovered by those chosen by the gods. Even in nineteenth-century America, this tradition is found in the story of Joseph Smith and the golden plates on which the Book of Mormon was written. Smith had performed a series of magical rituals in the woods of upstate New York and eventually an angel—Moroni—directed him to the spot where the plates were buried.

▼   ▼   ▼

We are taking some time to look at this myth—even before we begin to discuss the creation of human beings by Marduk or Enki—in order to focus on the peculiarity of the idea of authority and sovereignty and how it changed from a theocentric model to an anthropocentric model and how that switch has made it virtually impossible for human institutions to come to any kind of agreement on the nature of the Phenomenon. One of the sources for this interpretation of authority is a 2008 paper co-authored by Alexander Wendt and Raymond Duvall entitled "Sovereignty and the UFO."

The abstract that begins the paper sets out the problem clearly by stating that "Modern sovereignty is anthropocentric, constituted and organized by reference to human beings alone . . . enabling modern states to command loyalty and resources from their subjects . . ." However, "It has limits . . . which are brought clearly into view by the authoritative taboo on taking UFOs seriously . . . the puzzle is explained by the functional imperatives of anthropocentric sovereignty, which cannot decide a UFO exception to anthropocentrism while preserving the ability to make such a decision. The UFO can be 'known' only by not asking what it is."[7]

This is an important approach to the problem for it sets out in language and concepts borrowed from modern philosophers Jacques Derrida, Michel Foucault and Giorgio Agamben the essential problem at the core of the UFO discussion, which is that our human institutions—having seized authority (essentially, the Tablet of Destinies) from the gods—are now incapable of addressing the UFO issue. Since the Phenomenon demonstrates super-human characteristics that show it is not human-based or of human origin, our human institutions cannot say anything meaningful about it. The entire basis of human sovereignty is predicated on an anthropocentric view of the world, of reality. In other words, we will never have disclosure the way we understand it because that would involve our human authorities acknowledging another (higher, potentially more powerful) authority in the world. Paradoxically, if our human authorities *did* acknowledge the existence of another authority that would automatically undercut their ability to make such an acknowledgment!

Catch-22.

It was not always thus, however:

> Nevertheless, historically sovereignty was less anthropocentric. For millennia Nature and the gods were thought to have causal powers and subjectivities that enabled them to share sovereignty with humans, if not exercise dominion outright . . .
> In modernity God and Nature are excluded . . . [8]

This is what gives us the fundamental problem. We cannot assign the Phenomenon to the realm of religion since we do not respect religion as a source of knowledge about the world. If we refer to our Venn diagram (above) we can say that science has pushed religion from This World to the Other World. This World is the realm—now—of humans and of institutions based on what the authors call a metaphysics of anthropocentrism, and is no longer the realm of the gods. Humans are sovereign, and human sovereigns rule over other humans. Anything that calls this into question is a threat to modern ideas of rulership and to the loyalty of human subjects to human rulers. Again, even though the Earth is no longer the center of the universe, the anthropocentric—the "human-centric"—view of the world has replaced the theocentric, "god-centric" one. Humans are now the center of the universe and scientists and governments the sole arbiters of what is real; the entire structure of modern societies depends on it. In fact, the very word "real" is cognate with "royal": *reality is whatever the king says it is.*[9]

This is why the Phenomenon is unacceptable to both science and government. This is why there can be no such thing as a UFO. At least, that is why we are not able to define what it is for that would automatically challenge the sovereignty of our institutions. It would shift the focus away from us to Them, and we have no idea what the repercussions of that shift would be.

> For both science and the state, it seems, the UFO is not an "object" at all, but a *non*-object, something not just unidentified but unseen and thus ignored.[10]

What of theocratic societies? Even though modernity has dominated the worldview of societies in the West—in the United States, Australia and Europe, primarily—there are areas of the world that still emphasize the centrality of religion, such as regions in the Middle East, Asia, and Africa where economic development lags behind Western material achievements. It may be no coincidence that regions where economic development is slow are also regions where religious sentiments and a religious or "spiritual" worldview predominates. The association of matter with science, technology and anthropocentrism is reflective of Gnostic ideas of the corrupting influence of the material world. Yet, even in these cases, there seems to be a reluctance to accept or acknowledge the reality of UFOs. Is it possible that to do so would be to challenge certain dogmas?

In fact, we have evidence that even organized religion is willing to accept the idea of alien life. The Vatican recently came out with a statement that indicates a willingness to view the possibility of alien existence as not being at odds with Catholic belief. There has never been a problem with considering the possibility of alien life and alien contact in Islam, for instance, which generally has been more comfortable with science and scientific attitudes towards life on other worlds. Of course, this can be seen as a strategy to hedge one's bets, to maintain a degree of authority over the faithful by incorporating scientific and even pseudo-scientific scenarios and thereby extending sovereignty over all forms of knowledge.

In modern societies, however, the adherence to a scientific viewpoint as if it were an ideology or a dogma has effectively painted the skeptics into a corner. The omniscient god has been replaced by the omniscience of science and it cannot admit of knowledge that may be at odds with what is already known about the nature of the world and of reality. It cannot admit that the Other World has any kind of objective existence; instead, the Other World is a trash receptacle for "rejected knowledge." It is the domain of error, hallucinations, and the irrational.

To be sure, the Sumerians and the Babylonians entertained a similar viewpoint. For them, the Other World was the Underworld: the realm of the dead, of evil spirits, of all the ills that pester humanity and tend to inject an element of chaos into order and thus to challenge the supremacy of Marduk and the State. But to the Sumerians the Other World *did* exist; it was necessary to believe that it existed, otherwise there would be no need for divine authority. The existential threat posed by the Other World—the source of danger and hostility—made rulership and social organization necessary. The Sumerians did not deal with the irrational or the chaotic by wishing it away; they understood that eternal vigilance was required. Life was replete with dangers, from environmental catastrophes to sickness to defeat in war. All of these dangers emanated from the Other World, threatening to bring about a new chaos.

Marduk can provide us with a template for rulership in Babylon. Marduk was clearly a supernatural being with divine origins (either directly from Apsu, as indicated in some texts, or from Enki according to other texts) who fought against the hideous sea monsters created by Tiamat as well as against Tiamat herself. Thus he embodies both the very human attribute of a warrior as well as a spiritual force whose battles took place before Creation. By ordering the universe he becomes a secular ruler as well: the ultimate authority over the reality he created.

The kings of Babylon—Marduk's city—were also both sacred and secular rulers. It is important to point out that the division between sacred and secular did not exist as sharply in ancient Mesopotamia: a ruler was believed to possess authority over all realms of experience. Indeed, the word Babylon itself derives from a word of unknown origin—*Babili*—thought to mean "Gate of the Gods." It was a liminal space between this World and the Other World. In Hebrew, wordplay on the name gave it a different etymology: *Babel*, meaning "confusion," yet another ideological choice that can be said deliberately to characterize the religion of Babylon as a polytheistic confusion of gods as well as of tongues.

Research has shown that the word "Babylon" often was extended to identify other cities in Mesopotamia, almost as if the word was a title rather than a geographic location, but with the original site as being of prime importance. According to Babylonian sources, Babylon was the first city that Marduk created and for that reason alone it could be considered a "Gate of the Gods." The authority of Marduk extended from the temple enclosure at the top of the ziggurat—the point of tangence between this World and the Other World—down the seven levels to the surface of the Earth, and from there to the seven descending levels of the Underworld.

Some of this was retained in the Jewish tradition. The Tablet of Destinies became the Breastplate of Aaron with its associated Urim and Thummim: a divination system that is the subject of some academic discussion and controversy, but which seems to have been a simple method of casting lots, or dice. The temple enclosure at the top of the ziggurat became the Holy of Holies of Solomon's Temple; the New Year ritual of face-to-face contact with Marduk became the Yom Kippur ritual during which the High Priest was permitted to enter the Holy of Holies. Much of Genesis is an elaboration or rendition of material that has been found in the Sumerian and Babylonian texts: the Great Flood, Noah and his Ark, even the idea that the kings who lived before the Flood enjoyed enormous longevity, all are prefigured in Sumerian literature and belief. So, too, the famous Tower of Babel which is an obvious reference to the ziggurat at Babylon.

According to Genesis 11:1-9, this tower was built on the "plain of Shinar." Shinar is the Hebrew word for Sumer. The tower itself was intended to reach the heavens and so make a name for the people of Babylon. According to Genesis 11:5, "The Lord came down" and saw what the Babylonians were doing, complaining that ". . . nothing they plan to do will be impossible for them." For that reason alone, according to Genesis, God confused their languages so that they would not be able to cooperate. He made the people of the Earth divided so that they would not compete with him. That's it. That is the entire story. There is no mention of God punishing the Babylonians because of the sin of pride or anything else as this story has often been interpreted in Sunday school and Catholic confraternity classes. The only problem God had with the builders of the Tower is that they were able to do whatever they set their minds to do, and God could not allow that. So he "came down," a phrase that could be interpreted to mean "descended" as from the sky, or simply that he showed up in some fashion. In either case, he seems to have been absent during most of the construction of the Tower and only arrived when it seemed they were succeeding in their efforts. This continual absence of God from the Earth in Genesis is notable. The God of the Torah appears, issues edicts, then disappears. He reappears at moments when humanity seems on the brink of some major breakthrough, *and then stymies it.*

What we have here is an iteration of Genesis 3 where God has created two human beings, told them what to do and what not to do, and then disappears. Eve is told by the Serpent in the Garden that if she eats of the forbidden fruit she "shall be like God, knowing both good and evil." God reappears, and learns of the disobedience of his creations, which is really their desire to become the equal of God. God cannot tolerate that and removes them from the field of operation (the Garden of Eden) where their newly acquired knowledge would enable them to compete with him. And when the Babylonians demonstrate *their* capabilities by building a Tower, God once again intervenes to keep human beings from attaining divine status. Human beings in the Bible have slave-status vis-a-vis God, and it can be argued that this is an inheritance from Sumerian beliefs.

In Genesis, humans have the *power* to do what they want but they do not have the *authority* to do what they want. The authority is God's alone. The biblical version represents a monotheistic authority, and the *Enuma Elish* (among other Sumerian, Akkadian and Babylonian epics) represents a polytheistic authority, but in essence they are identical. For all practical intents and purposes, authority—whether in a monotheistic or polytheistic context—remains the most important attribute of a god. Any human authority must derive from that divine authority; it must be bestowed by a god upon a human, and can be taken away as quickly as it was given.

Thus, too, the Phenomenon represents a challenge to human, anthropocentric authority. The sight of a vessel flying through the air would not be ignored or ridiculed in a society where authority emanates from a divine source. The paranormal in general is consistent with religious documents, rituals, beliefs, and experiences (one of the reasons it is ridiculed by science: guilt by association). Visions, dreams, apparitions, and supernatural phenomena create a space for the experience of the UFO.

In a modern society, however, where a constantly evolving scientific paradigm determines the contours of reality, these phenomena are judged to be non-existent. They lack reality. They are not "real" in any kind of scientific sense and thus there is no need and no attempt to explain them, analyze them, critique them, because to do so would be to undercut the very foundational model of science. They do not represent phenomena that can be tested, predicted, or measured even though they are known to leave physical traces. Theories of UFOs are not, in scientific parlance, falsifiable.

In a modern society, where a dominant political model of human rulership demands loyalty and obedience from humans to humans, these phenomena are ignored. To do otherwise would be to undermine the very foundational model of human sovereignty. As we are told in 1 Samuel 15:23, witchcraft—the domain of antinomian religious practice and belief—is the equivalent of rebellion against the state. It is an old equivalency—one that obtained in ancient Sumer as it did in ancient Palestine—and it holds to this day. As there was no science (as a recognized discipline separate from a religious or spiritual context) in those days then methods of science that were not part or supportive of the prevailing attitude also would be considered witchcraft and treasonous. *Vide* Galileo.

Ironically, the only way any of this makes any sense at all is if we remember the near universal understanding that all authority ultimately derives from the fact that human beings are not autonomous; they are not self-creating but merely self-replicating, like von Neumann's universal constructor, (more about this later in Book Three). Humans are a product of something else, some other factor, that the ancient sources tell us were the gods.

One Million Years a Slave . . .

About two million years ago we saw the evolution of the first humans from their primate ancestors. This was *Homo habilis* or "handy man": possibly the first tool-making primate ancestor, and is the subject of some controversy since not everyone agrees that this species was a hominid. Then, about one million years ago, *Homo erectus* ("upright man") made its appearance. Its origins are equally controversial, with some paleontologists suggesting that *Homo erectus* arose first in Africa and then spread to parts of Europe, Central Asia, China and Indonesia, with other experts suggesting that Asia was the point of origin. It has even been proposed that some of the earlier extinct hominid species—such as *Homo habilis*—should properly be classified under the *Homo erectus* category.

Eventually, however, *Homo neanderthalensis* made its appearance. Earliest forms seem to date to about 500,000 years ago with the fully formed Neanderthals only about 250,000 years ago, and then became extinct about 40,000 years ago. *Homo sapiens*—of which we humans are a subspecies, known as *Homo sapiens sapiens*—arose around 200,000 years ago and co-existed for a time with the Neanderthals until the latter became extinct. We share an almost identical genetic heritage with Neanderthals, with only about a 0.12 percent difference between us.

According to genetic and paleontological evidence, therefore, the origins of humanity are broadly (if somewhat incompletely) known. We can trace our lineage back two million years (at the most liberal guess) to one million years (the conservative estimate) in time. At the very latest, we can use the 200,000- to 250,000-year estimate of the appearance of both the Neanderthals and *Homo sapiens* as our point of origin. It really depends on what one means by the term "human" and especially "modern human."

There are questions of genetics, of course. There are also questions of consciousness. Did a larger brain size indicate greater intelligence, and did greater intelligence contribute to the development of consciousness? These are still controversial issues today, especially as it has become obvious that not everyone agrees on the definitions of terms like "intelligence" and "consciousness." If we do not know how to define consciousness, then we are unable to determine when it first developed. We are not even confident enough to state unequivocally that animals have consciousness; we are certainly nowhere near accepting that plants have consciousness, much less the elements themselves that make up human beings, animals, and plants.

As we will see in a later chapter, the genetic origins of consciousness are now being hotly debated. Without going into that controversy in detail, we can ask a few basic questions here since they go to the problem of the origin of human beings, and this problem relates directly to the question of the Phenomenon itself.

First of all, let us assume that the nature of humanity and of human beings presupposes the existence of consciousness. We feel we are different from every other creature on Earth because of our consciousness. We are introspective beings who wonder about things like life, death, the afterlife, etc., and we do not perceive the same concerns taking place among other members of the animal kingdom or the plant kingdom. We also seek to control our environment in ways that no other species, genus, family, etc., has done. We build cities, cars, rocket ships. We design intelligent networks. Create fiber optic cables and irrigation systems. We do not see other species doing the same. Ergo, we believe, it is our consciousness that sets us apart from the rest of Creation. We calculate, measure, manufacture. We create works of art, music and literature. We do not see the same activity elsewhere among other creatures.

Yet, everything that lives has genetic material. There is now a growing body of evidence to suggest that the genetic code originated elsewhere than on our planet, and that it was somehow "seeded" onto the Earth: either accidentally, due to a meteor strike or some other natural mechanism, or deliberately. This begs the question: is "life" the same as "consciousness"?

Again, we do not agree on definitions of either of these terms so it becomes an exercise in pure speculation to come down on one side or another of this discussion. Instead, let us look for an answer—or at least trace evidence—in the historical record.

The *Enuma Elish* tells us that humans were created from the blood of the slain Kingu. Remember that Kingu was appointed by Tiamat to be commander of the forces fighting Marduk, and he was also the god to whom Tiamat gave the Tablet of Destinies. Thus Kingu represents the enemy of Marduk, and it is from Kingu's blood that humanity is formed. Humanity, therefore, traces its origins to a weak, illegitimate, and defeated spiritual force. The purpose behind our creation was to work as slaves for the gods who are therefore relieved of any hard labor and who are thus assured of a continuous supply of food and drink. This is exactly as it was stated in the ancient Sumerian and Babylonian texts; the elaborate theories of Zecharia Sitchin and others that humans were developed to engage in mining operations, etc., simply are not necessary in order to make the point that the Sumerians believed that humans were created to work for the gods. It's an unnecessary complication for which there is tenuous (at best) evidence.

The cosmos itself was created from the body of Tiamat. Her bones, blood, and organs were used in various ways to complete the organization of the external world. The blood of her consort Kingu was used to create human beings. Thus, humans are not of the same "genetic line" as the rest of the universe, although their origin is due to divine energy (the will of Marduk) and divine substance (the blood of Kingu). It should be noted that the god who actually does the work of creating the humans, once told to do so by Marduk, is the Sumerian god Enki.

Enki was the lord of fresh water, of artifice, magic and the ability to manipulate matter. His temple was known as the *E-abzu*, with its clear reference to the *abzu* or abyss. In other versions of the Creation story, Enki takes a more central role in the creation of human beings but always winds up using the blood of another god (in one case, using clay mixed with blood), and always for the purpose of using human beings as slave labor.

Eventually, however, the slaves revolt. When that happens, the gods decide to wipe them off the face of the Earth with a Great Flood. Deluge stories and legends are known virtually worldwide, and that may be due to the memory of a single flood that took place in the distant past or, more likely, refers to various floods occurring in different parts of the world at different times. Many people living today have had experience of floods; they are a relatively common occurrence so there is no need to look for a single cataclysmic event that affected the entire globe. What is a common denominator between the Babylonian account and that in Genesis, however, is that both floods are the result of the gods or God becoming angry with human beings: in other words, humans brought the flood upon themselves. The other common denominator is the idea of a special human who was spared, along with his family.

This story may be taken two different ways. In the first place, we may be seeing an ancient attempt to describe natural selection. This is the essential theme of Darwinism, which states that some species die out and others survive due to their inability (or ability, respectively) to adapt to environmental forces. When some humans survive a catastrophe, a cataclysmic event like a tsunami or other natural disaster, we may describe them (and their offspring) as having been the product of natural selection. Of course, this is not exactly true. In Darwinian terms, this process takes place over long periods of time and is not the result of a single event taking place over weeks or even months but rather adaptations and mutations of the species that survive under hostile conditions or develop as strategies appropriate to their particular environment. It may be that what the authors of Genesis were describing was a memory of other hominids becoming extinct, leaving only *Homo sapiens sapiens* alive to tell the tale.

The other way of interpreting these stories is to suggest that the selection was not natural at all, but deliberate. In fact, in Genesis, God specifically tells Noah to build the Ark. Once again, authority comes from a divine source; the power to build the Ark, however, is human. Humans, after all, are still slave labor, even as late as Genesis 6.

Before that happens, however, there is a bit of stage-setting. One of the prominent theories of Sitchin and others both before and after him concerns an enigmatic reference in Genesis 6, which is the chapter of the Bible dealing with Noah and the Flood. It opens with a reference to "the sons of God and the daughters of men" and then on to the Nephilim. This is one of those areas that has been exploited by Sitchin, von Däniken, and others who have promoted the "ancient alien" theory, because it is one of the strangest episodes of the Bible. It is actually one of the places in the Bible that could be used to support an ancient alien theory if it hadn't been misinterpreted by Sitchin and others who went overboard in their zeal to prove their own ideas: Sitchin, about gold mining operations, the Annunaki, and a "twelfth planet," and von Däniken with his insistence that the "sons of god" were aliens engaged in artificial insemination of human beings. If we separate these fanciful notions from the texts themselves, we can see that there are grounds for supposing that this small section of the Bible—barely a paragraph—gives us the best indication yet that something truly weird took place in dim pre-history and that the memory of it was fresh enough (or traumatic enough) that civilizations around the world retained some elements of it in their creation epics.

There Were Giants in Those Days . . .

The New International Version (NIV) of the King James Bible begins Genesis 6 this way:

> When men began to increase in number on the earth and daughters were born to them, the sons of God saw that the daughters of men were beautiful and they married any of them they chose. Then the Lord said: "My Spirit will not contend with man forever, for he is mortal, his days will be a hundred and twenty years."
>
> The Nephilim were on the earth in those days—and also afterward—when the sons of God went to the daughters of men and had children by them. They were the heroes of old, men of renown.

There are a great many mysteries tied up in these few lines. Who were the "sons of God"? Who were the Nephilim? What is the meaning behind all the non sequiturs in these two paragraphs? Why does God interject his statement about his Spirit not contending with man forever between the two descriptions of the sons of God and the daughters of men? And if the children of their union gave rise to "heroes of old, men of renown" (if we are supposed to read the sequence that way) then why does the author of Genesis complain that men were wicked, in the paragraph that follows?

> The Lord saw how great man's wickedness on the earth had become, and that every inclination of the thoughts of his heart was only evil all the time. The Lord was grieved that he had made man on the earth, and his heart was filled with pain. So the Lord said, "I will wipe mankind, whom I have created, from the face of the earth . . ."

Obviously, we're missing something.

The "sons of God" were supposed by many commentators to mean "angels" or some type of supernatural being. Later research indicates that the term "son of God" was used to refer—not to angels, but—to a kind of royalty. To kings, or to high-ranking ministers. That indicates a social disconnect between men and "sons of God." The opening sentence says that the sons of God saw that the daughters of men were beautiful. Thus, the sons of God were not men in any kind of normal sense but they had bodies that functioned as those of humans.

The Hebrew term used in Genesis 6 is *bene elohim*. It is a term that has been used in different ways to mean slightly different things. David Penchansky, a professor of theology, writes that bene elohim refers to a "divine council": a group of beings who serve as ministers to God. They either have human bodies or can assume human bodies in order to function on the earth. When they do assume human bodies, it seems that they are able to propagate through sexual intercourse with human women (the bene elohim are exclusively male, as the account in Genesis 6 suggests) and produce offspring from this union. (In common UFO parlance, these might be considered "hybrids".)

▼   ▼   ▼

Enter the Nephilim.

Sitchin wants to translate this word as "Fallen," i.e., fallen angels, after the Hebrew word *naphal*. That was a popular interpretation among biblical scholars for a while, but it since has been demonstrated to be faulty and not consistent with rules of Hebrew grammar. It was actually a loan word from Aramaic, *naphiyla* that had been Hebraicized and given its plural form *nephilim*.[11] For some reason, the translators of the King James Bible kept the word Nephilim in their version without translating it themselves. There was confusion—or perhaps astonishment—over the real meaning of Nephilim which is probably what kept it in a transliterated form.

Nephilim, according to scholars of Hebrew and Aramaic, can only mean one thing: *giants*.

This was not a word used in a general sense, metaphorically referring to someone of great personal stature or charm. The word refers specifically to beings of extraordinary size. Genesis tells us the Nephilim were around in those days *and afterwards*. The implication being *after the*

Flood. The further implication is that they are no longer around, for they were only around "in those days—and also afterward," but evidently not now. It may be the Nephilim who are described as "men of renown."

There is also the repetition of the phrase "on the earth." It appears four times in those six verses. Not to be coy, but where else would man be but "on the earth"? Were there men elsewhere in Creation? Was it only those "on the earth" who had become troublesome in the eyes of their Creator? Otherwise why was the repetition—a form of emphasis—necessary?

And where did the Nephilim, the giants, come from?

One reading—the most commonly accepted—links the offspring of the sons of God with the daughters of men with the Nephilim: in other words, when the two mated these monstrous forms were created. That would indicate that the sons of God were not part of the same gene pool as the men and the daughters of men.

In other words, the kings of the Earth, the *bene elohim*—the royalty, those holding an authority from God—were not human themselves. The human men were the slave race, as indicated above. Their daughters were part of that same race, obviously. The mating of the daughters with the *bene elohim* produced giants: genetic abnormalities.

Briefly, we considered whether what was being described was some kind of Neanderthal. Were the Nephilim—or their progenitors—Neanderthals? Unfortunately, for many reasons, that can't be true. For one thing the Neanderthals were considerably shorter than *Homo sapiens sapiens.* They may have appeared more brutish, perhaps, but they were a good head or more shorter than their hominid siblings. Definitely not giants.

Is there some clarification to be found in the earlier, Babylonian, sources?

Fortunately, there is.

One of the most famous personalities from Babylonian literature is Gilgamesh, the hero of the Gilgamesh Epic and intimately associated with the Babylonian Flood stories. In order to fully understand the idea of the Nephilim, then, we have to go back to earlier, non-Jewish, sources that had a more finely articulated history of this period and that help to clarify matters even as they present a startling alternative view. This includes the Gilgamesh material and those ideas associated with it, including the *apkallu* or *abgal*.

The *apkallu* (to use the Babylonian term; *abgal* to use the Sumerian) were seven wise beings, seven sages, who existed before and after the Flood. They were responsible for teaching the newly created human beings the benefits of civilization. These were not humans, but divine or quasi-divine beings who could assume human form when necessary or desired. In other words, they were the template for the biblical idea of the bene elohim.

The most controversial version of the story of the apkallu comes from Berossus who was a priest of Bel Marduk in the third century BCE. According to his account, one of the apkallu—named by him Oannes[12]—would rise from the sea (presumably the Persian Gulf) each morning and teach the people what they needed to know about agriculture, astronomy, etc., and then would return to the sea at dusk. He would not eat or drink anything on land, and he dressed in an unusual suit that resembled a fish with scales. He wore a head covering which looked like the head of a fish, and his clothing resembled that of a fish except for the fact that he walked on two feet.

Eventually other apkallu followed the first, and they would remain in contact with the human beings for some time, at least up to the Flood. They were said to have their origins in the Abzu, the Abyss of Enki, and they were sent back there when they had angered Marduk who viewed them as having corrupted the human race. This, however, did not take place before the apkallu began to mate with the human women and produce offspring that were part human, part apkallu. These elements—taken from a variety of Sumerian and Babylonian sources—clearly anticipate Genesis 6.

If biblical scholars such as Michael Heiser and David Penchansky are correct then there is a strong Mesopotamian tradition of a council of normally invisible beings—the apkallu, the abgal, or the bene elohim—who appeared on Earth as advisors to the human race and who eventually mated with human women, creating monstrous offspring in the process. The biblical account is clearly a somewhat sanitized version of the Mesopotamian original, given a monotheistic spin. If we subtract the ideological changes from the story we can see a very strong indication of a general belief that there was non-human contact with humans at some point in pre-history. This non-human contact is not merely a euphemism for God or angels, because the words that are used for these terms in the Bible (*el, elohim*, YHVH, *malakim*, etc.) are quite different from those used to characterize the apkallu (bene elohim). These were beings of an order quite different from humans. Normally invisible, they would assume human form to interact with human beings and even produce offspring from sexual intercourse with human women.

Gilgamesh was one of these.

According to what is generally known as the *Epic of Gilgamesh*—in reality a collection of cuneiform tablets containing narratives about the hero—Gilgamesh is "two thirds divine" and "one third human."[13] He obtained knowledge of the world from before the Flood by speaking with the Babylonian version of Noah and learning how he became immortal.

Now largely believed to have been a historical figure (circa 2600 BCE), Gilgamesh had entered the realm of supernatural legend long before his presumed reign. He had rejected the advances of the goddess Inanna, for instance, and fought the demon Humwawa. In one of the tablets containing the *Epic of Gilgamesh*—found in the ancient city of Ugarit (Ras Shamra)[14]—we learn that Gilgamesh is a giant. He is described as being eleven cubits tall and four cubits wide. If each cubit is about 20 inches long, eleven cubits would mean Gilgamesh was 220 inches, or about 18 feet tall and 80 inches or 6 feet wide. His being a giant, and two thirds divine and one third human would make him one of the Nephilim.

The king of Uruk, Gilgamesh was as much a culture hero as a godling. We may say of him that he was a "man of renown." He certainly was a warrior, but his most famous exploits are those involving supernatural beings such as the demon Humwawa whose face is usually portrayed as a mass of entrails. In fact, the name of Gilgamesh in cuneiform is often preceded by the *dingir* symbol, which is usually reserved as the identifier of a god.

He is someone who straddles the line between apkallu and human, a being who was conceived of a human woman but whose paternity remained a mystery. In fact, according to the single legend concerning his birth, his mother—surrounded by armed guards to ensure that she would not be molested and lose her virginity illegitimately—became pregnant with him even as she was under lock and key. Alarmed, the guards took the baby and threw him from the top of a tower to be dashed on the rocks below, but an eagle arrived just in time and settled the baby Gilgamesh safely onto the Earth.

Thus, without addressing it in so many words, the origin of Gilgamesh is that of a "daughter of men" and a supernatural father, an apkallu, and he developed into a giant who was two thirds divine. This is in keeping with the biblical tradition as well as the Mesopotamian one. In fact, according to the *Epic of Gilgamesh*, the apkallu designed and built the walls of his city, Uruk.

The apkallu themselves are often depicted as having human bodies but the heads of birds (similar to the type of iconography we find in Egyptian illustrations of Thoth,

Horus, etc.). Mesopotamian bas reliefs depict them as a type of griffin with four wings as well as bird heads, indicating their capability of flight. In fact, it is probably these huge reliefs that were found in Babylon and elsewhere that gave rise to the famous vision of Ezekiel 1 (a favorite of ancient alien theorists) in which the cherubim make their appearance as having four wings and four faces: those of a man, an eagle, an ox, and a lion. As Ezekiel had this vision in Babylon it could be said that the local artwork influenced its content.

A further connection of Gilgamesh to the Flood story—that starts with the discussion of the "sons of God" and the Nephilim—is his visit to Uta-napishtim (Sumerian: Zi-u-sudra), the Babylonian Noah. It was from this personality that Gilgamesh learned of the ways of the world before the Flood, for Uta-napishtim survived the deluge in an Ark with his family, plants, and small animals. Uta-napishtim had saved the seed of humanity and for this was rewarded by the gods with immortality.

There is no space to go into this story in full, but it is mentioned to show that Genesis 6 has its origins in Babylonian religion which itself had its origins in Sumer.

To summarize the data:

The universe was created as a result of a struggle between invisible forces—characterized as fresh water and salt water which gave birth to gods who turned on their parents and used the body of one to create Reality and the blood of another to create Humanity. Thus humans are a product of one of the defeated gods. In some versions of this narrative clay from the earth is used as a medium, mixed with divine blood. The purpose of the creation of human beings was as slave labor for the gods.

Also present in the heavens were intermediary forces—a kind of assembly or council known as the apkallu in Babylon, the abgal in Sumer, and the bene elohim among the Jews—who were in charge of the situation on Earth. These beings were able to assume human form but were not human beings. They copulated with human women (there does not seem to have been female apkallu to mate with human men) and the result of their intercourse was a generation of Nephilim, or giants.

For reasons which are not entirely clear, God or the gods decided to destroy the human race by sending a Flood. However, they decided that some genetic material could be salvaged and that resulted in the stories of Noah (in the Bible) and Uta-napishtim (in Babylonia).

At some point after the Flood the gods (or their emissaries, the apkallu/abgal/bene elohim) disappeared. Suddenly, human beings were on their own again. The loss of contact with these supernatural beings was lamented in the Sumerian hymns and incantations. Rituals evolved to bring them back, at least temporarily, using sacrifices and sacred marriages. The sacrifices recall the blood that was used to manufacture human beings, acknowledging that our blood is not really ours, but theirs. The sacred marriages recall the sons of God mating with the daughters of men. Using these two mechanisms we were reminding the apkallu of this ancient ur-event in human history, reenacting over and over—through thousands of years—that initial trauma; or perhaps we have been working through that trauma ourselves, using ritual to try to understand what had happened and what our purpose is on this planet.

▼   ▼   ▼

These stories were well-known throughout the Levant as late as the first and second centuries CE. They were inextricable elements of the culture going back thousands of years. There are New Testament references to these same bene elohim, and to the seven sages—the apkallu—who were punished for having intercourse with human women and producing their hybrid offspring. They are not called apkallu in the New Testament, of course, but are cited in 2 Peter and Jude as "angels who sinned" and who were imprisoned in Tartarus (in the Septuagint Greek, usually just translated as "hell" in English translations) or in "deep gloom." Tartarus, of course, is the underworld Abyss of Greek mythology and the place where the Titans—who themselves "sinned" against the older gods and were overthrown—were imprisoned.

This idea of a war in the heavens which led to the creation of the world and of human beings, from divine substances, is familiar throughout the West. Roman and Greek mythology abound in examples, and the same theme would find itself represented in the Abrahamic literature as well. Like the stories about a Great Flood, these "myths" reflect knowledge of, or experience of, an actual event or events.

The idea that human beings were created as slaves or play-things of the gods also is present everywhere we look. And the "high strangeness" element—that some type of being with a supernatural origin mated with human women in the distant past to create hybrid offspring—is also very familiar to Sumerians, Akkadians, Babylonians, and Jews, not to mention Romans and Greeks.

The figure of Prometheus in Greek mythology shares many elements in common with the Near Eastern narratives. He is regarded as the divine being (actually one of the Titans) who brought wisdom, arts, and learning to human beings and who stood up for humans in opposition to Zeus (and was punished for it). Prometheus most famously stole fire from the gods and gave it to humanity. He was then chained to a rock for eternity, having his liver eaten every day by an eagle only to have it grow back at night, until he was rescued by Hercules. In his anger, Zeus fashioned Pandora—the "first woman"—out of clay and sent her to live with men as a punishment. These various motifs are seen by scholars as iterations of the Mesopotamian originals.

▼   ▼   ▼

During our research program for these books one of us—Tom—had an interview with a high-ranking member of the military establishment. After going over some of the material we had collected, he suggested—quite strongly—that we take a closer look at Greek mythology. This would have seemed like an odd suggestion were it not for the fact that the Sumerian and Babylonian "myths" were the original forms of the Greek myths and we were already hard at work on deconstructing those elements to uncover evidence of contact.

There could, of course, be another reason for this emphasis on Greece. After all, the Greek alphabet is a necessary tool for mathematicians and scientists. Greek mythology finds itself over-represented in naming protocols for planetary bodies such as the moon of Jupiter: Titan.

Hellenistic influence on Israel was profound in the Second Temple period and led to the revolt of the Maccabees in 167 BCE (memorialized in the Jewish festival of Hanukkah, a religious holiday that did not exist previously). The revolt was begun when King Antiochus of Syria ordered an altar to Zeus built in the Temple of Solomon in 168 BCE and outlawed Judaism. He even ordered pigs be sacrificed there, which is about as bad as it gets. At the same time there was a serious conflict between traditionalist Jews and Hellenizing Jews, but in 165 BCE the Temple was retaken and rededicated (which is the meaning of the word Hanukkah).

The Hellenistic period also gave us Hermetism: that form of esotericism founded upon the Emerald Tablet of Hermes and its associated documents concerning alchemy, philosophy, cosmology, and the secret sciences. The Emerald Tablet opens with the line, "That which is below is like that which is above, and that which is above is like that which is below . . ." as good as any a rephrasing of the statement in Genesis that has God in Heaven creating human beings on Earth in his "image and likeness." The connection between "above" and "below" was absolute and concrete in the minds of the ancients, the one a mirror of the other, a reflection of the event that they memorialized in their literature, arts, and sciences. Today, however, that mirror—that window into the Other World—has been covered up as humanity "sits shiva" for the death of their dreams.

There are many other possible avenues to pursue in Greek culture, history and religion, but at heart the most important event revealed by *all* of these sources—reiterated time and again, so often that we ignore it as background noise—is the contact between human beings and Something Other. There is even the constant reminder that humans were deliberately created by this Other to serve a specific purpose and when we developed our own intentions and our own goals (however misguided they might have seemed to be, eating of the forbidden fruit or building a Tower to Heaven) this Other tried to stop us and even to destroy us.

To regard this as purely a religious, or spiritual, or even superstitious belief is an ideological choice that has nothing to do with the evidence. The basic elements of the narrative you have read so far use ancient texts to demonstrate that a real event took place at some point in human history, an event that has caused ripples in every aspect of human endeavor. Ancient peoples did not have a vocabulary for what they experienced or remembered; as a matter of fact—if we judge by the vast majority of UFO sightings, abduction reports, and other aspects of the Phenomenon— *neither do we.*

That is why we have chosen to graph this discussion in some basic ways, such as describing the Phenomenon as representing a liminal space between this World and a putative Other World, a world that can be approached through what the Sumerians and Babylonians called a "Gate of the Gods." This Gate may be actual in some sense we do not yet understand (but which was understood by the ancients, and perhaps by those who have had some measure of "contact" in our times), or it may be virtual. Most likely it is a virtual Gate that acts as a mental or psychological construct that mimes the action of the actual Gate . . . but we are getting ahead of ourselves.

When the impulse to redefine the religious experience that began with polytheism to a more rigid and doctrinaire monotheism began to take shape the knowledge that was inherited from the Sumerians and Babylonians—knowledge that claimed direct contact with non-human actors— was rejected in favor of the new paradigm. Monotheism became the lens through which the past was re-imagined as a time when human beings were savage, uneducated, and gullible and therefore their accomplishments and histories were devalued as the product of ignorance. Forms of ancient knowledge—although by now fragmentary and incomplete—were preserved in the world of superstition and esotericism, the domain of the secret society and the sorcerer. It has only been within the last hundred years or so that much of this data is being recovered and reevaluated, leading academics and scholars to take a fresh look at the origins of western monotheism and to realize that many polytheistic concepts that derived from this contact survived within the Abrahamic religions in various forms.

The great advantage of monotheism over polytheism, however, was its potential as a unifying factor against the gods. In Sumer—as in later Babylonia—each city-state had its own patron deity. The same would hold true throughout the ancient world in Europe as well as in the Middle East. This meant there were political divisions between human societies that could be exploited to keep humans weak and unable to resist. A monotheistic worldview would unite these city-states under the banner of one patron deity with all of the citizens of those states united as well in a common political and theological system. Perhaps this was more than a purely ideological challenge of the monotheists against the rest of the world; perhaps it was an act of defiance against the "sons of God," and probably with reason.

As we have mentioned already—but which bears repeating as we go along and the material becomes more and more challenging—what we are proposing here and in the chapters and the books that follow is a scientific resolution to the old problem of gods, demons, good and evil, religion, magic, and all the impedimenta of the pre-scientific age. What we are proposing is that the ancients were recording an event or events that had a profound effect on human development, on civilization, on technology, an event that is irrupting into human consciousness and experience today in several different forms. The ancients framed this event in terminology that we read as religious or superstitious today, but these designations did not exist in ancient times. Even the concept of "religion" is relatively new. In Sumer, religion and science were not separate categories. There was only one approach to knowledge and it involved the whole mind.

The difficulty we face today in understanding or even defining the Phenomenon is the result of this fragmentation of knowledge which has led to the consigning of an entire realm of human experience to an ontological dustbin. Science cannot explain the Phenomenon so scientists refuse to see it; similarly, there is no space for spirituality in any form within science so scientists consider it a form of mental illness: a scientific-sounding euphemism that replaces such emotionally loaded terms as "superstition" and "ignorance," but which is just as loaded in other ways.

What may be required, then, is a re-engineering of the way we look at reality and of our responsibility towards other human beings in the face of our inability to deal with the Phenomenon in any other way.

This is the Sumerian cuneiform sign for "heaven" or "god," *dingir*. It is an eight-pointed star and is often found inscribed near a figure on a stele or a cylinder seal to identify the figure's divine status. The association of a "star" with "god" thus goes back to one of the earliest civilizations on Earth.

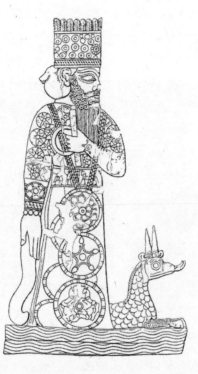

The Sumerian deity Marduk, shown here with the dragon Mushussu, often interpreted as the "red snake" or the "snake of splendor." Notice the multiple *dingir* symbols on Marduk's arms. The association of serpents and dragons with divinity would be retained as late as the Gnostics of the first few centuries CE in their depictions of Abraxas and the Demiurge. It would also be found in Greek mythology in sculptures and bas-reliefs of the Gorgons.

Marduk defeating Tiamat, from whose blood he would create the human race. Notice the weapons in his hands, which are similar to the *vajra* or *dorje* – thunderbolt – device familiar to Tibetan Buddhism; also, the seemingly anomalous "wristwatch" on Marduk's wrist which bears the eight-rayed star design, or *dingir*.

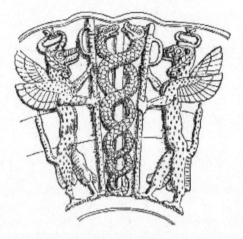

The Sumerian god Ningizzida (circa 2000 BCE). Two serpents coiled around a central shaft represent Ningizzida, the god of the Underworld; he is braced by two gryphons. This symbol will find itself represented as the caduceus in Greek mythology, and as the symbol of Kundalini in Indian yoga. As we will see in Book Three, it is also the symbol of the double helix of the DNA molecule.

A reconstruction of the Ziggurat at Ur. The ziggurat was a stepped pyramid with a temple at the very top where the priest-king would meet with the goddess on the New Year.

STAIRWAY TO HEAVEN

The volume of reporting is related to many things. We know that reports of this kind go back to biblical times.

−Major General John A. Samford,
Director of Intelligence, United States Air Force,
in a press statement made on July 29, 1952,
concerning the UFO sightings over
Washington, DC, earlier that month.

He had a dream in which he saw a stairway resting on the earth, with its top reaching to heaven, and the angels of God were ascending and descending on it.

−Genesis 28:12

A stairway to heaven shall be laid down for him, that he may ascend to heaven thereon.

−The Pyramid Texts, Egypt, circa 2300 BCE

N THE PREVIOUS CHAPTER WE LOOKED AT ONE ASPECT OF the ancient Mesopotamian belief system which included definite references to the creation of human beings by "the god" to function as slave labor, and the actions taken by God or the gods to keep human beings from controlling their own destinies; to deprive them of authority and of what modern anthropology would term "agency." We examined the persistent theme of the "sons of God" who mated with human women to produce a hybrid race of giants. These concepts are taken directly from the texts themselves (the *Enuma Elish*, the *Epic of Gilgamesh*, the *Atra Hasis*, and even the Bible) and are not invented or embellished by us. We suggest they be considered as documentary evidence for an event that transpired in pre-history that we are calling—for sake of a better word—contact.

There is one other persistent theme in these traditions and that is celestial flight. In other words, buried deep within the stories, legends and rituals of the ancient Near and Middle East (as well as in cultures worldwide) there is the belief that humans—at least, some humans—are able to travel to the stars themselves, essentially reversing the original flow of life from a celestial "creator" to the Earth by going up a ladder or a pole from the Earth to make direct contact with the Other World. To standard, mainstream religious leaders, such an attempt would be anathema. It would be tantamount to building another Tower of Babel, and we all know how that turned out.

This type of practice was banned by most religious traditions in the West, and is reserved for persons of special characteristics or abilities in the East (such as shamans). Within Judaism, this practice is known as *merkava* mysticism or *hekhalot* mysticism which is intimately connected with Kabbalah, although probably much older than the earliest Kabbalistic texts. *Merkava* (or "chariot") mysticism derives from the first chapter of the Book of Ezekiel in which the Jewish prophet describes a vision while in exile in Babylon of a chariot descending from the heavens. This vision became the model for deliberate attempts to reach the celestial realms through the use of meditation and other methods for achieving altered mental states.

In other words, there was an understanding—and this may be critical for a serious comprehension of, and approach to, the Phenomenon—that altered states of consciousness were necessary in order to acquire contact with the "sons of God." While certain physical aspects of the initial, pre-historic contact were mimed—magic circles with lamps, incense smoke, strange sounds and arcane languages—in order actually to "fly" heavenwards a modification of the mind's normal processes was required. If we consider these practices (and they can be found all over the world) as evidence of a real experience that took place millennia ago, then we should be able to apply them to the current problems we encounter in designing systems for space travel, especially the type of travel that would approach the speed of light . . . or exceed it. Those problems may involve propulsion issues, psychological conditioning for extended periods of space flight in zero gravity, etc., or there may be a more immediate application.

In our conversations with such controversial figures as Bob Lazar and others who wish to remain anonymous we learned that a major obstacle facing the scientists who wanted to re-engineer captured alien spacecraft was the control issue. Without coming down on one side or the other on the validity of the "captured alien spacecraft" meme we wish to point out that the difficulty in imagining how to control a vehicle that had no visible means of steering or velocity control—or even an ignition switch—may be an allusion to how the ancients experienced these same "vehicles." What is largely missing from the Sumerian, Babylonian, biblical and other accounts of sightings and contact is the technology involved in controlling the vehicles. No one seems to be steering or otherwise physically manipulating these craft. This leads us inevitably to wonder if they are controlled by some sort of mental mechanism that we do not yet understand.

In order to get a grip on this admittedly bizarre concept, we will take the time in this chapter to see how celestial flight was understood and what common denominators may be found across cultures and eras. Then, in the following chapter we will look at what happens when celestial flight breaks down, when contact is lost, and a civilization goes crazy as a consequence.

Jacob's Ladder

Genesis 28:20-22 tells us the story of Jacob, the Ladder, and the "gate of heaven." The story is familiar to readers of the Bible as the scene in which Jacob falls asleep with his head on a stone as a pillow, and has a dream of a ladder reaching to the heavens with angels "ascending and descending," the voice of God from above telling Jacob that the land will belong to him and to his descendants.

There is a great deal more to this story, and it has implications for our analysis of the Phenomenon. As it stands, there is virtually no other scenario like it in the older Sumerian or Babylonian sources which are, however, replete with tales of gods and goddesses traveling to the underworld down a series of seven levels or steps. There are also many instances of shamanic figures rising to the heavens up a pole or a tree to make contact with non-terrestrial forces. Thus it would be of use to analyze the biblical scenario in light of what we know from these other sources.

On its face, the story of Jacob is a mixture of Jewish history and supernatural events. Jacob was the son of Isaac and the father of Joseph: he of the coat of many colors who became an advisor—a dream interpreter—to the Pharaoh of Egypt. It was Jacob who wrestled with an angel; Jacob who was renamed by that angel "Israel."

The corporeality of angels in the Old Testament is an issue to which we will return a little later on in this chapter; it's one of the most jarring characteristics of angelology in the Bible for it goes against the popular notion of what an angel is supposed to be, but is much closer to what we discussed in the previous chapter concerning the "sons of God" and the apkallu.

For now, though, we will focus on the episode of Jacob's Ladder.

One of the problems of biblical interpretation is the way in which we project backwards in time, analyzing biblical events as if they were superstitious stories intended for the entertainment of children or the gullible. The other approach is to take every word as the literal truth. Often the reality may be found somewhere in between these two polar opposites.

In the case of Jacob's Ladder we have a real event that has been obscured by so much reverse interpretation that it has lost much of its impact. Since we do not, as moderns, credit any story that includes a tangible, visible connection between this world and an alleged or putative "other" world, the entire scenario must be an invention, or even a deliberate deception.

Jacob was the twin brother of Esau. The Bible is very clear on this point. They were both contending with each other in their mother's womb. Jacob wanted to be the first born of the two and held onto his twin's foot as if to drag him back into the womb. That is how he got his name "Jacob" which means "one that takes by the heel." Esau, on the other hand, was born with a full head of hair and complete bodily hair as if he was already an adult; thus his name comes from the Hebrew for "finished."

As the first born, Esau was destined—according to tradition—to inherit his father's wealth. His father was Isaac, the son of Abraham, the founder of what we now call the Abrahamic religions: Judaism, Christianity and Islam. Isaac was nearly slain by his father when God ordered Abraham to sacrifice Isaac as a demonstration of his loyalty. At the last moment, an angel stayed Abraham's hand and the life of Isaac was spared.

Thus we have the appearance of an angel who is the intermediary between Abraham and God and who convinces Abraham that the original commandment of God that Abraham sacrifice Isaac is not to be followed. This must be seen in context.

At the time, child sacrifice was not uncommon, even among the Jews. The biblical references to Moloch should be proof enough of this, but there is much other archaeological and textual evidence that child sacrifice was common in the Middle East at the time of the Abraham and Isaac episode. As we will see in the next chapter, child sacrifice was also common in the New World: among the Olmecs, Toltecs, Aztecs and Incas. It may be too harsh or outlandish a critique, but what if child sacrifice performed a type of genetic selection function? Certainly, with every child that was killed before they reached puberty and could have children of their own their genetic substance was effectively removed from the gene pool. If, as the ancient texts suggest, human beings were created as a race of workers for the pleasure of the gods, then they would have a vested interest in weeding out those children who were unique enough genetically to pose a threat to the status quo. As it turned out, Isaac was one of those children because from him issued the entire kingdom of Israel: a nation that would eventually turn to monotheism as a rejection of the plurality of the gods and of their henchmen, the "sons of God."

Isaac eventually married Rebekah—a woman from Abraham's native Mesopotamia—and it was Rebekah who bore Esau and Jacob. Prior to her marriage to Isaac she was blessed by her family with the words, "Our sister, may you increase to thousands upon thousands, may your offspring possess the gates of their enemies" (Genesis 24:60). This same blessing is still used today in traditional Jewish weddings.

Esau and Jacob, her sons, were of completely different dispositions. Esau was a hunter, an active man, rough, who—against his parents' wishes—married local Canaanite women who were idol worshippers. The implication is that Esau was also an idol worshipper.

Jacob, on the other hand, was a quiet, introspective sort who spent most of his time in study and pious reflection. His mother decided that when the time came it would be Jacob who should inherit his father's legacy and not Esau whom she saw as irresponsible and uncaring about his birthright. Thus the famous episode in which she dressed Jacob in the skins of animals—to imitate Esau's hairy body—and sent him in to the nearly blind Isaac to get his father's blessing instead of Esau. Thus, Jacob's inheritance of his father's birthright was the result of an elaborate deception on his part and his mother's. It is often justified on the grounds that Esau was not a fit candidate due to his lack of respect for his parents, his tradition, and his people.

Jacob fled his home ahead of Esau who was filled with rage at the deception and wanted to kill Jacob. Jacob was making his way to the home of his mother's family when he fell asleep on the stone and had his famous vision:

> He had a dream in which he saw a stairway resting on the earth, with its top reaching to heaven, and the angels of God were ascending and descending on it. There above it stood the Lord and he said: "I am the Lord, the God of your father Abraham and the God of Isaac. I will give you and your

descendants the land on which you are lying. Your descendants will be like the dust of the earth, and you will spread out to the west and to the east, to the north and to the south. All peoples on earth will be blessed through you and your offspring. I am with you and will watch over you wherever you go, and I will bring you back to this land. I will not leave you until I have done what I have promised you."

When Jacob awoke from his sleep he thought, "Surely the Lord is in this place, and I was not aware of it." He was afraid and said, "How awesome is this place![1] This is none other than the house of God; this is the gate of heaven." (Genesis 28:12-17)

There is again the emphasis on reproduction, of the spreading of seed throughout the world from a single source that we found in the blessing of Rebekah, Jacob's mother. The theme of reproduction—of massive reproduction to cover the earth—is a constant one in Genesis. Indeed, it is from the same root that we get both words: Genesis and genetics.

But what is of equal importance is the idea that there are angels ascending and descending the "stairway to heaven." First, the angels are mentioned as ascending the stairway. From where are they ascending? We must assume it is from the Earth. The implication is of ongoing two-way traffic of these angels between the Earth and Heaven, which is consistent with our concept of the bene elohim or the apkallu as remaining influential in terrestrial affairs long after the Flood.

The second important concept is that of a technology— the stairway, or actually ladder (in Hebrew, *sulam*)—that permits two-way traffic between the Earth and the heavens. While we learned in the previous chapter of the "sons of God" mating with the daughters of men and producing giant offspring (the Nephilim), we now have a glimpse of how they "landed" on the Earth.

Another theme that can be noticed is the repetition of the phrase "on Earth" or "of the Earth," with the Earth referenced as a place distinct from other places. God is clearly not on the Earth or of the Earth. Neither are the angels. The Earth is a laboratory, a test site (there are many "tests" of humans by God throughout the Bible), an object viewed as if from afar by beings who are not part of its environment or its context. These are beings who have no personal stake—no real investment—in the outcome of the planet and who can casually destroy entire cities and races of people in a fit of pique. In the case of Jacob (as with his mother Rebekah) there is a definite plan in mind. Their offspring will reproduce unhindered. This is a promise of God to Jacob, who is now assured that God will assist him in his every endeavor to cover the Earth with his offspring. That seems to be the sum total of God's interest in Jacob: one involving territory and offspring. Basically, God is promising Jacob *Lebensraum*, and tells Jacob he will not leave him until he has done what he has promised: implying that when he has accomplished this goal, he will leave.

When he awakens, Jacob anoints the stone he had been sleeping on. Pouring oil over a stone is an ancient Middle Eastern custom. The stones were often meteoric stones, and the name for such a stone is *baetylus*. This is a Greek word that actually derives from the biblical account, for Jacob names his stone Beth-El, which means "house of God." *Bethel* is the original form of *baetylus*.

There are many such traditional stones in the region. The Dome of the Rock in Jerusalem is built over one such stone (believed by some to be the spot where Abraham was supposed to sacrifice Isaac). Even the Ka'aba of Mecca is said to be of meteoric origin and its position as an object of worship predates Islam by many years. In fact, the Prophet Muhammad—a member of the Quraysh tribe that was in charge of the Ka'aba when it was a pagan shrine—cleared out its 360 pagan idols and re-dedicated it as sacred to Allah, thus reinforcing the idea that the meteoric rock was a symbol of the connection between this world and the Other World. If we consider that the Dome of the Rock is another important Islamic shrine, then there is a kind of axis of polarity between the two stones; this is emphasized by the role that the Dome of the Rock plays in what is known as the *Mi'raj*: the flight of the Prophet Muhammad up the seven heavens on a mysterious beast called the Buraq that seemed to be a kind of flying horse.

In this instance, the Prophet is sleeping at the Ka'aba in Mecca when the Buraq appears along with the angel Gabriel. The Prophet mounts the Buraq and is taken immediately to "the farthest mosque" which most agree is a reference to the Al-Aqsa mosque in Jerusalem (where the Dome of the Rock is located). From there, he ascends to the heavens to meet all the prophets who have gone before him. He is then returned to Mecca on the Buraq. It was during this flight that the Prophet is told he must tell his followers to pray five times a day.

Thus we once again have the presence of an angel, a flight to the heavens, and a commandment from God. The presence of a meteoric stone in both the accounts of the Mi'raj and Jacob's Ladder may be significant.

Because of the presence of magnetite in meteoric rocks, a piece of these rocks could be used in the manufacture of compass needles. Not only did these rocks "descend from the heavens," but they also pointed to the North Star. They played the role of cosmic messengers as well as navigation devices. This dual function was not lost on, for instance, the ancient Egyptians who used a device made of magnetite in the shape of the Big Dipper (called the "thigh of Set" in the "opening of the mouth" ceremony in which the mummified Pharaoh would be re-animated by this magical implement.

"Opening of the mouth" ceremonies were used in Babylon, as well, in order to animate statues of the gods. The idea was that the opening of the mouth and nostrils of either the mummy (Egypt) or the statue (Babylon) would enable the figure to breathe and speak. The association of the meteor from the heavens, its magnetic quality, and immortality as well as the association with the gods, is a concept that reinforces our suggestion that life on this planet was believed to come directly from a supernatural source and did not arise spontaneously on Earth. The involvement of "angels" in subsequent iterations of this theme further emphasizes this connection. In fact, the very concept of the apkallu survived into the immediate pre-Islamic period among the Bedouin in the form of the abgal (the Sumerian word for apkallu) as a local deity that was half-human, half-fish: a clear reference to the Oannes of Berossus. The primary deity of the pre-Islamic Ka'aba was Hubal, a god that was brought to Mecca by the Quraysh tribe from Mesopotamia and specifically from what is now Iraq. The function of the Prophet Muhammad, therefore, was similar to that of the Jewish prophets: to eliminate polytheism and replace it with a pure form of monotheism. Like the early Jewish prophets, however, that did not mean that the Arabs stopped believing in the existence of other gods: they simply relegated them to subordinate status, or classified them as demonic. This is not strictly speaking monotheism, but henotheism: a kind of preference of one god over all others.

Henotheism is a reasonable outgrowth of polytheism in the face of evidence that multiple supernatural forces exist. Whereas the polytheist may feel inclined or compelled to worship a multitude of deities, the henotheist recognizes that the other deities exist but has committed to the worship of only one above all others. The henotheist may even redefine the other gods as angels, demons, or—in the case of the Bedouin—*jinn*.

Although mentioned several times in the Qur'an, and especially Surat 72 which is entitled *Surat al-jinn*, information concerning the jinn is known to predate Islam by hundreds of years. The jinn are invisible to human eyes but are just as corporeal and can interact with humans, according to Arab and later Islamic belief. Like humans, they also have free will and thus some have converted to Islam (as the above-mentioned Surat attests). A book published about the jinn by the Foundation for Islamic Knowledge[2] in the United States, for instance, devotes more than half of its chapters to a discussion of Shaitan and Iblis (two names used for the ruler of the jinn). The idea is that the jinn more closely represent demons than humans and although some have converted to Islam it is still forbidden for human beings to converse with them in any way. They are said to have been created from "smokeless fire," molded into human shape, and can sometimes take the form of serpents and other unclean animals.

Iblis, the king of the jinn, was ordered to kneel before Adam to show respect for the created human but he refused and for this reason was cast out of Heaven. This is a story similar to that told by the Yezidi who claim that Shaitan committed this sin out of love for God for he could not imagine bowing before anyone but God. The Yezidi believe that Shaitan will be the first of God's creatures to return to Heaven and when he does the Yezidi will follow him.

This conflict between God and the creatures of God is therefore a common theme throughout much western religion and culture. The "sons of God" who mated with human women created such a problem on Earth with their quasi-hominid offspring that God felt the only solution was to destroy the world with a Flood and start over. This conflict did not end, for when Jacob begins his return home after his long sojourn away he is confronted with another angel, and this time is forced to wrestle with it.

Jacob and the Angel

The account given in Genesis 32: 22-32 is as ambiguous as the story of Jacob's Ladder. It again involves a supernatural being who becomes involved in Jacob's life for no immediately discernible reason.

At this point in the narrative, Jacob is moving with his family and worldly possessions back to his home. He finds himself alone on one side of a river at which point—and à propos of nothing—someone comes out and begins to wrestle with him:

> So Jacob was left alone, and a man wrestled with him till daybreak. When the man saw that he could not overpower him, he touched the socket of Jacob's hip so that his hip was wrenched as he wrestled with the man. Then the man said, "Let me go, for it is daybreak."
>
> But Jacob replied, "I will not let you go unless you bless me."
>
> The man asked him, "What is your name?"

"Jacob," he answered.

Then the man said, "Your name will no longer be Jacob, but Israel, because you have struggled with God and with men and have overcome."

Jacob said, "Please tell me your name."

But he replied, "Why do you ask my name?" Then he blessed him there.

So Jacob called the place Peniel, saying, "It is because I saw God face to face, and yet my life was spared."

Whereas much that happens in the life of Jacob is clear narrative, with an identifiable cause-and-effect, linear storyline involving human beings with discernible motives and reactions—Isaac, Rebekah, Esau, etc.—when it comes to the supernatural elements they are out of context and seem unrelated to the general narrative.

In this case, someone comes out of nowhere and begins wrestling with Jacob. He is described as "a man" repeatedly, which indicates that as far as Jacob was concerned (at least, initially) he was dealing with a being that was fully corporeal. In order to defeat Jacob, the "man" had to exert a little Vulcan on Jacob's hip, thereby dislocating it and causing Jacob to limp forever after.[3] There is no rational explanation for any of this, in spite of the fact that most of the rest of the story is entirely rational or at least credible to a degree, allowing for exaggeration or poetic license here and there. This episode, however, is dropped into the narrative flow and there must be a reason.

When Jacob leaves his home for the first time he has his encounter with the Ladder. When he returns home, he has his encounter with the "man." He later identifies this man as an angel or actually as God for the word Peniel means "face of God." He claims he has seen God face to face, and survived. His name is then changed to Israel, which means "he strives with God."

So he has gone from Beth-el—or "house of God"—to Peni-el, or "face of God." Beth-el was "the gate of heaven," a Jewish version perhaps of *Bab-ili* or Babylon: the gate of God. We still do not understand the motive of the "man" for wrestling with Jacob, unless it was a poetic device to show Jacob's strength and qualifications for being the sire of an entire people, the people of Israel who were named after him. In any event, it is entirely possible that the segue from "man" to "God" is predicated on the idea that the being with whom he wrestled was an angel (as it is often interpreted), one of the "sons of God" that we have encountered already.

In a paper published in *Glossolalia*, Yale scholar Kathryn Pocalyco offers an interpretation of the story based on anatomy.[4] She suggests that the injury suffered by Jacob was not caused by a blow to the hip socket itself (since that part of the anatomy is inaccessible externally) but by a blow to the heel which would then transfer the force of the blow to the hip joint, thereby dislocating it. There is a pleasing symmetry about this solution, for Jacob's name refers to "the heel" since he was grasping Esau's heel at birth. The wounding of Jacob's heel is a kind of karmic balancing of his debt to Esau for having deceived him and robbed him of his birthright. Thus this story sets up the next day's meeting between Jacob and Esau: a meeting that Jacob had been dreading and which is the reason he stayed behind after sending his family on ahead. This solution has the advantage of rescuing this bizarre episode and placing it (somewhat precariously) in the narrative flow. It still does not, however, explain how it came about in the first place.

The conclusion of this specific episode is vitally import-ant for it introduces the name and concept of Israel and reaffirms Jacob's role as the father of the tribes even as he is going to meet his brother, whom he wronged. Jacob asks for the man's/angel's name and is instead rewarded with the question, "Why do you ask my name?" Jacob then jumps to the conclusion that the man was actually God himself.

Jacob's experience of the Ladder took place at night, a time of darkness and sensory deprivation; so did his encounter with the wrestler. In both cases, Jacob was on the verge of a major change in his life. In the first instance, he was fleeing his twin brother Esau. In the second, he was going to meet him. If we were to interpret this story from a shamanic perspective, we would find that enough elements are present to support at least a second look from that point of view: a ladder reaching to heaven, an experience of God, visions of angelic forces, a struggle and physical maiming as a result, leading to direct contact with the divine.

Another telling detail is the limp.

As described elsewhere, this characteristic has a long pedigree in esoteric symbolism.[5] A limp is associated with the shamanic Chinese practice known as the "Pace of Yu."

This practice involves the shaman "walking" on the stars of the Big Dipper—known as the Great Chariot, itself a suggestive reference—on the way to the celestial pole which is the throne of immortality. The Pace of Yu is based on the story of the shaman known as Yu who became ill and developed a limp after a particularly strenuous ordeal. The Great Chariot itself is often represented in China as personified by a Goddess of the Chariot, sometimes depicted as riding in a chariot being drawn by seven boars (representing the seven visible stars of the Dipper). The stars of the Dipper are understood by the Daoist practitioners of the Pace of Yu as a ladder leading from the Earth to the heavens and ultimately to the pole around which the entire cosmos turns. The Dipper in this instance serves as a gateway to Heaven, just as Jacob's Bethel.

In the second degree initiation of Blue Lodge Freemasonry, there is a similar ritual. The basic symbol of this degree is the Winding Staircase, which can be interpreted as a version of Jacob's Ladder. The initiate is expected to walk in an exaggerated fashion around the area of the temple, to the four cardinal points of the compass. Remember that God tells Jacob that his seed will extend to the four quarters of the world. The Masonic temple represents Solomon's Temple in Jerusalem which is where, of course, one finds the sacred rock of the Temple Mount, the site from which the Prophet Muhammad performed his famous flight to the heavens.

What these brief references are meant to convey is a constellation of themes cognate to celestial flight, experiences of angels or other supernatural beings, and the resulting "initiation" of a human being in contact with those forces. The themes we discussed as mentioned in Genesis concerning Jacob are those that can be found in a number of other places around the world, in different cultural contexts, and as such interpreted differently in each culture but nonetheless demonstrating many elements in common.

If we entertain the idea that the angels so often mentioned in the Bible are related to—if not identical with—the strange beings we studied in the previous chapter that were corporeal, could appear and disappear at will, could even cohabit with human beings, and who acted as representatives of the Creator on Earth, then we have biblical as well as Babylonian precedence for the modern concept of alien contact. Perhaps the reader can see why the term "alien" is so problematic in this context, for it is a contemporary label loosely applied to an ancient experience. The ancients were actually much more explicit in the way they described these beings, as intermediaries between us and the forces that created us. Themes such as reproduction, blood, seed, and subservience to authority were interpreted in purely religious contexts as civilization changed from the Sumerian model to more recent versions and as that transformation was taking place the original information was reframed as fables and superstition.

There is, however, a *technology* at work here. This technology is of the very essence of those machines we have seen in the skies above our world, a technology that involves both the visible, material world as well as the unseen world of our ancestors. It involves the first examples of celestial flight.

Ezekiel's Vision

Jacob had his vision of a ladder and angels ascending and descending, with God standing at the top of the ladder looking down from Heaven. Not to be outdone, Ezekiel had a vision of God descending from Heaven, not down the steps of a ladder but in his own chariot. This vision inspired a mystical practice that is known as *merkava* or "chariot" mysticism, the basic method of which involves ascending the (usually seven) levels of Heaven to reach God. Merkava mysticism is also known as "the *Descent* to the Chariot," for reasons that are not very clear at all.

As mentioned above, Chinese mystics call the Big Dipper the Great Chariot or sometimes the Golden Chariot. To the Babylonians, the Dipper was known as the Great Wagon (MAR.GID.DA). The ancient Greeks referred to it as the *amaxa* or "chariot." To the Arabs, it represented a funeral bier (which nonetheless retains the sense of a wheeled vehicle). The Dipper is an asterism, part of the Ursa Major constellation, and it circles around the Pole Star. To find it, one looks north.

On July 31, 593 BCE, by the banks of the River Chabar in Babylon, the prophet Ezekiel had a vision. According to his account, he was looking due north when the vision occurred:

I looked, and I saw a windstorm coming out of the north—an immense cloud with flashing lightning and surrounded by brilliant light. The center of the fire looked like glowing metal, and in the fire was what looked like four living creatures. In appearance their form was that of a man, but each of them had four faces and four wings. Their legs were straight; their feet were like those of a calf and gleamed like burnished bronze. Under their wings on their four sides they had the hands of a man. All four of them had faces and wings, and their wings touched one another. Each one went straight ahead; they did not turn as they moved. (Ezekiel 1:4-9)

There is much more, but for now we can begin by making some preliminary observations. The vision—as we shall see—is eventually described as a chariot. That it was seen in the north may be a further indication that Ezekiel is referencing the Dipper. The four living creatures, along with their wheels within wheels (Ezekiel 1:15-21) can be said to represent the four stars of the Dipper that were commonly used to represent the rectangular body of the Chariot.

The description of the "four living creatures" is remarkably similar to something that Ezekiel would have seen in Babylonian captivity: the huge bas-reliefs of the apkallu: half-bird, half-man creatures with four wings and the hands of a man that were a common feature of the architecture. The vision seems to be a coded reference to something taking place that could only be explained by inserting clues recognizable to those Jews living in Babylon during the Exile. That would include not only the architecture, religion and customs of the Babylonians but also the architecture and religion of the Jews, especially where Solomon's Temple—the First Temple—was concerned: the temple that was sacked and destroyed by the Babylonians.

Ezekiel's Vision became the focus of intense scrutiny, not only among the Jews but also to later generations of Christian Kabbalists as well. In fact, study of Ezekiel's Vision was prohibited to all but the most advanced Jewish scholars and even then it was not to be discussed with others.[6] It seemed to hold a key to understanding how human beings could make contact with the divine. It is this idea of Ezekiel's Vision as an instruction manual for alien contact that brings it to our attention as we try to strip away some of the purely denominational or ideological accretions—not easy to do when discussing the vision of a pious Jewish priest—to find the technology beneath the surface. Ezekiel stood at the crossroads of Jewish and Babylonian—and therefore monotheist and polytheist—traditions and his vision reflects those elements even as it introduces new ones. If we look at them through the lens of what we know of the Phenomenon, however, a spectrum of clarity is introduced that transcends traditional (mainstream religious) ideas of God, gods, angels, and even prophecy.

So much has been written in an attempt to decode Ezekiel's Vision from the point of view of UFOs, ancient astronauts, and the like, that to reprise all of those theories here would be counterproductive. That Ezekiel is talking about an event so bizarre that he is at a loss for adequate language to describe it is obvious. It reads like the first science fiction story ever written, with a being in an impossible vehicle that seems humanoid but not completely, descending to the Earth and (in Ezekiel 8) virtually abducting the Jewish priest. These quasi-human beings are already familiar to us from the Mesopotamian and biblical contexts, and it is odd that there has been so little focus on this aspect of the religious traditions that have contributed to the ancient alien hypothesis. Rather than speculate about the *Annunaki* or the *Igigi* and theories concerning mining operations and missing planets, a more fruitful approach might be to contemplate the persistence of historical accounts involving beings that have hominid features and characteristics, but which are perceived as non-human in some particular way, as these experiences are inescapable and still occur with regularity down to modern times. It is necessary to ask why there is a near-global pattern of contact with these non-human beings and why there is a similar pattern in global creation literature of the "manufacture" of human beings by other forces: something that is not obvious, but which is somehow inferred and taken for granted. A look at Ezekiel's Vision and how it has influenced beliefs and practices concerning non-human contact may provide us with an answer to some of these questions.

The language of Ezekiel's Vision is not ordinary language; that is, it cannot refer to ordinary events or commonplace occurrences. Although the words can be defined and the grammar is normal, the way in which the vision is described defies rational interpretation. This is a phenomenon that we find throughout mystical literature, not only in the Middle East but in India and other parts of Asia as well. In fact, this tendency to write in what seems to be a deliberately obscure fashion is also found in European alchemical texts. In India, this is called *sandhya-bhasa*, or "intentional language," sometimes translated as "twilight language." The later merkava texts—the *Ma'aseh Merkava* or "Works of the Chariot"—that use Ezekiel's Vision as their template, are written in a similarly obscure prose. This seems to represent an effort to transcend normal rules of language as the experiences they describe transcend normal experience. This is, of course, part of the problem in the modern era when eyewitness statements concerning the Phenomenon are ignored by academics and scientists because of their wooly language and "other worldly" terminology; this results in statements being dismissed because the eyewitnesses or experiencers are not scientifically trained observers (which is most people) and thus an entire body of evidence is ignored because the human observers did not go to the same schools or get the same degrees as the scientists.

Ezekiel's Vision begins as noted above, with the appearance of a kind of vehicle (he does not use the word chariot—*merkava*—until much later) that has an unknown or unidentified propulsion system. According to Ezekiel, the faces on the four sides of the body of the vehicle face in the direction of travel. These four beings are referred to as the *hayyot*, or "living creatures." Below the four beings with the four wings and the four faces—man, lion, ox, and eagle—can be seen the wheels, or the *ophanim*. They are described as a "wheel inside of a wheel" and seem to run along the outside or perimeter of the vehicle.

Sitting on a throne made of sapphire in the vehicle is "the likeness of a man." He seems to be the one who controls the vehicle. That he is sitting is relevant, for the hayyot do not sit; they are specifically described (in the excerpt above) as having legs that are "straight." The implication is that they do not bend at the knee. This may refer to a tradition that angels do not have knee joints; i.e., their legs are always straight, they are always standing at the ready. In later works of the merkava mystics we learn that Metatron— sometimes identified with the prophet Enoch—sits on a throne before the inner sanctum where God is to be found. The fact that he is sitting causes some mystics to mistake him for God, since angels cannot sit down; at least, not in the presence of God. These hayyot, therefore, may be angels; or they may simply be the struts that support the vehicle as they have the appearance of "burnished bronze."

Above the head of the man in the throne is "an expanse, sparkling like ice" and the sound of the rushing of the wings like "the roar of rushing waters." The man himself is fire from the waist down, and glowing metal from the waist up. He is surrounded by a radiance like a rainbow.

A voice comes out of the vehicle, directed at Ezekiel, who falls down before it. He is told to take a scroll that is handed to him, to eat it, and then to repeat its words to the people of Israel in exile. It is a prophecy full of dire warnings.

Here we must stop for a moment to digest the above scene. It is obvious why recent generations have interpreted the above as describing a UFO. To be fair, it *is* a flying object, and it *is* unidentified (at least insofar as the description does not make a lot of sense). Yet, the most revealing part of this Vision may be the prophecy itself.

It is a truism that virtually every experiencer who has had contact with the Phenomenon in a close encounter with an "alien" has come away with a dire prophecy for humanity. These involve end-of-the-world scenarios, environmental disasters, and world wars. The aliens never arrive with a message of praise and congratulations for their human hosts. Instead, they show up in supernatural vehicles with awesome technological capabilities, appear humanoid— with the "likeness of a man"—and then deliver their hideous prophecies, and leave.

What we wish to point out here is the relative consistency of this experience from ancient, biblical times to the present. The fact that Ezekiel eventually will be abducted by the being in the chariot is further evidence that we are dealing with a single, unitary complex of ideas and images that is repeated endlessly from era to era, from culture to culture.

We don't wish to go into great detail on the nature of Ezekiel's Vision and his subsequent prophecies concerning Israel, the Temple, etc., as they are specific to his time and place. There are Mesopotamian motifs here and there in the narrative, which is only to be expected, and we may wonder if the "likeness of a man" seen by Ezekiel is related to the humanoid figures that appeared to the Sumerians and gave them knowledge and wisdom, but a comprehensive analysis of the data is beyond the scope of this work. What is of relevance to us now, though, is the fact that this document became the model for one of the earliest forms of Jewish mysticism, older even than the Kabbalah itself. As mentioned above, this practice was considered so dangerous that only the most pious and most learned were permitted to study it and even then they could only discuss their findings—and, presumably, their experiences—with one other person and no more.

In fact, this practice can be found at the heart of some of the Jewish messianic movements that spread throughout Eastern Europe beginning with Shabbtai Zvi and, later, Jacob Frank. These movements also influenced the development of some modern esoteric societies and occult orders, and one can trace the evolution of much New Age spirituality and organizations back to that day on the banks of the Chabar River in Babylon when a priest in exile looked up to the northern sky and saw a chariot flying down to confront him, challenge him, and carry him away.

Works of the Chariot

First and foremost, the chariot as seen by Ezekiel, and which appears again to carry the prophet Elijah to heaven (2 Kings 2:11), is a vehicle. It is, in all of these cases, a flying object. It implies transportation, but of a celestial character. The chariot is a device for traveling to heaven. There is even a chariot *in* the heavens, the asterism known to us as the Big Dipper but to many other cultures worldwide as a wagon (Charles' Wain, in England) or a funeral bier (among the Bedouin), but mostly as a chariot. Thus, the Works of the Chariot involve celestial travel.

Celestial travel is usually understood in this context to mean a spiritual experience, something that occurs as the result of prayer, meditation, and contemplation. That is not the whole story, however, for not all prayer and meditation leads to the experience of "descending to the Chariot." Rather, what we are discussing is a very specific form of practice with a unique methodology, and it is based on an actual event. It is as if a cult had developed around Kenneth Arnold's sighting in 1947, or the experience of George Adamski a few years later, by people who would imitate as closely as possible the circumstances of the original sightings. In other words, it is not to be understood as "spiritual" in the generally accepted sense of the term, but as a technology whose aim is to duplicate the original experience by adopting many of its features. Like the cargo cults we discussed at the very start of this study, what is happening is what appears to be a ritual on the one hand and a practical exercise on the other.

A modern person stumbling upon the ersatz airstrip, bamboo air traffic control tower, and coconut headphones of a South Pacific cargo cult would recognize the scene immediately even though it was not a real airport with any hope of witnessing an aircraft land. The same is true of the Works of the Chariot. They represent a simulacrum of a genuine event; and while it is possible that a Pacific Islander who belonged to a cargo cult could have a spiritual experience—i.e., an experience that alters his or her consciousness due to the intense focus on the rituals of the cult—so, too, does the "descender to the Chariot." It is this experience that may be the key to understanding the mechanism of the Chariot: the controls, propulsion, and navigation systems. This is because our linear approach to this problem is inadequate (so far) and does not take into account the different intellectual and psychological structures of the visitors who created them. In order to get close to an appreciation of the Phenomenon we need to "see" differently, and practices such as the Works of the Chariot enable us to do that.

Second, the descender to the Chariot works alone. In fact, the Rabbinic tradition demands it. Ezekiel was alone by the river when he had his experience. This sense of isolation from society—at least for the duration of the experience—is as much a modern phenomenon as it is biblical. Modern experiencers (who, by the way, are almost always accidental descenders) rarely have their experiences in groups. Abductions usually involve only one person at a time, usually at night. This isolation indicates that the experience is an internal one, whether generated internally or due to a selection process which we do not understand as yet.

Third, the practitioner must be psychologically, intellectually and spiritually prepared. There is a very real reason for this, and that is because this type of experience can unhinge a person who is unstable mentally or emotionally. This is a caveat from virtually every mystical practice in the world, from Tantric yoga to ceremonial magic and including, of course, the Works of the Chariot. The intensity required by this practice is such that one easily can become deluded, prey to hallucinations and feelings of grandeur on the one hand or utter despair on the other. It also requires a great degree of persistence and as the practice is a solitary one that means the practitioner must rely on his or her own inner resources.

Fourth, in addition to being alone during the practice one must engage in a certain amount of sensory deprivation. The less distraction from the external world, the better. In terms a modern would understand this is to create a communication pathway between one's conscious mind and one's unconscious mind, something that is difficult to accomplish when external stimuli demand attention.

Fifth, there is a requirement for a certain amount of intellectual preparation as well. The Works of the Chariot involve ascending up seven celestial levels. Each level is guarded by its own angelic force which means one must know the seal, name, and password appropriate to each level and be possessed of enough strength of mind to remember which is which and to use them at the appropriate time even in the face of strange and unsettling images, sounds, smells and other experiences. This is a distinct inheritance from the Sumerian and Babylonian narrative of the descent of Inanna/Ishtar into the Underworld, down a series of seven steps at each one of which she is required to remove a jewel or article of clothing. These objects can be understood as passwords or as token payment for each level. Inanna will—at the end of the narrative—ascend up those same seven levels and re-acquire the objects she traded at each level. Inanna's goal at the end of her descent was to stand before the throne of Ereshkigal, the Goddess of the Underworld, who then kills her and leaves her body hanging on a hook on the wall for three days before she is resurrected. The goal of a descender to the Chariot is to rise up the seven heavens and stand before the throne of God, which is located among seven palaces—or *hekhalot*—on the topmost level. The descender will then return to the world after this experience a changed human being.

Parallels to shamanic celestial flights are numerous. This idea of flying upwards to the heavens and becoming transformed by the experience is nearly universal. In some cases, one returns with enhanced powers and capabilities; in others, with greater spiritual insight. In almost all cases, one comes back with experiences that are indescribable in words but which paradoxically enable the experiencer to become an asset to the community, even though the community has no means of comprehending the experience itself.

The descender is transported through the air, either gradually or in an instant. There are different degrees of the experience of being transported, as the Prophet Muhammad had during the Mir'aj. In the New Testament, Paul recounts a similar experience of rising up to the third heaven in what must be one of the earliest recorded instances anywhere of a Ma'aseh Merkava. The difference between these ancient accounts and modern ones is that the ancient accounts are almost always framed in a religious or spiritual context, whereas modern accounts are often secular or spiritual in a non-denominational sense. Thus, it may be time to strip this experience—and the technologies for achieving it— from their spiritual and religious trappings and understand it for what it is: contact.

The technology we are calling the Works of the Chariot opens a door between This World and the Other World, in other words, the realm of the Phenomenon. It enables a human being to instigate contact consciously and deliberately, rather than having it happen only accidentally if at all. Yes, we know this sounds very New Age-y and mystical, but that is not our intention. Rather such a perverse reaction is the side effect of so much poor contextualization of the experience itself. We need to remove this process from its cultural and religious ornamentation in order to see it for what it really is.

Of course, the original descenders to the Chariot worked within a strong cultural, linguistic and religious context. That was their ground, the intellectual and emotional basis from which they moved from the Earth to the heavens. But for the purposes of this study and for our overall project we want to demonstrate that these ideas are not merely the superstitious yearnings of the disenfranchised or the crackpot beliefs of a few religious fanatics, or the loopy tantras of the credulous and hopeful. These ideas have been marginalized largely due to their association with specific religions or with religion in general, but the practice that underlies them is solid. As we will see in Book Two, governments and armies around the world understood this and put some of it to work in the twentieth century, even trying to improve upon these practices and eliminate the human factor as much as possible (which turned out to be counter-productive, as could have been predicted).

There were, however, cultures in which these practices were either non-existent or deliberately suppressed; where the individual was sacrificed for the good of the community and thus where a descender to the Chariot would not have survived. When this particular, unique approach that links the world of human beings with the world of the "sons of God" is missing or forgotten—but the initial event, the contact between humans and the Other is remembered, including the memory of the creation of human beings from the blood of slain gods—then what transpires is a desperate attempt to revisit that ancient contact by any means necessary.

A painting by William Blake, showing Jacob's Ladder.

Quelle: Deutsche Fotothek

From a European alchemical text, the seven stages to perfection marked by a temple at the top of the "magic mountain" with sun and moon, surrounded by the Zodiac and the four elements. Compare this image with that of the Ziggurat at Ur from the previous chapter.

Like Jacob, and like the European alchemists, the Prophet Muhammad also experienced celestial ascent. This painting depicts the seven levels or seven heavens which are a standard image of Jewish *merkava* and *hekhalot* mysticism as well.

An illustration for Dante's *Paradiso*, Canto III, by Gustav Doré, showing the "highest heaven."

The Vision of Ezekiel, which has given rise to so much speculation concerning Biblical UFOs. The winged creatures, the circles within circles, the divine communication in the form of a warning, with a burning city in the background, are themes that are encountered in the accounts of alien abductees and also in those of mystics and shamans.

The body of a Pharaoh being prepared for "celestial ascent" during the ceremony known as Opening the Mouth. The mummification rituals involved a device, an adze, in the shape of the Big Dipper, which symbolized the Egyptian asterism "the Thigh of Set", which is represented by the thigh of a calf. These complex images represent a detailed itinerary for the Pharaoh's resurrection among the stars. Some have speculated that the mummy in its sarcophagus could represent, for modern people, an astronaut in his space-suit.

THE CULT OF THE DEAD GODS

But where are we to go now?

We are ordinary people,

we are subject to death and destruction,
    we are mortals;

allow us then to die,

let us perish now,

since our gods are already dead.[1]

> – a plea of the Aztecs to their
> Spanish conquerors

Did I get this straight?

Do you want me here

As I struggle through each and every year.

And all these demons, they keep me up all night.

> – "Up All Night," Tom DeLonge (2011)

IN PREVIOUS CHAPTERS WE LOOKED AT THE SUMERIAN and Egyptian cultures and noted their focus on astronomy, astrology and rituals that connect ideas of death and the afterlife with otherworldly journeys. It is obvious that humanity has been concerned with the stars and with a belief that the gods originated from there, and that humans have the stars as their final destination. That may not be enough to convince many people of an extraterrestrial influence on humanity, however. Just because we yearn towards the stars does not mean that we have been visited by beings from the stars. Even though our cultural documents insist on an extraterrestrial (i.e., divine, angelic, etc.) explanation for human technological and intellectual capabilities and are replete with examples of humans and gods interacting, the lack of hard evidence to support this claim—evidence that can have no other, alternate explanation—will leave many unimpressed.

Yet, there is evidence of another type that is often overlooked, even by the most vocal proponents of an ancient alien hypothesis, and that is the evidence of a deeper connection between human beings and non-human intelligences than can be found in anomalies of architecture or even in the myths themselves.

That is the psycho-biological connection.

As an example, a characteristic of religions and cultures worldwide is the phenomenon of blood sacrifice, even human sacrifice. Variations of human sacrifice have surfaced everywhere from ancient Israel to the British Isles, from the South Pacific to India, and in the Americas. Cultures and civilizations that had no discernible contact with each other developed forms of human sacrifice for many of the same reasons. The idea that the gods want or need or deserve human blood obviously is a characteristic of the human condition that so far has evaded serious study.

While there have been books written on the subject of human sacrifice they have mostly been focused on the practices and on the justifications for it in diverse cultures rather than representing an attempt to understand it from a larger perspective. What complicates matters is that these same studies tend to associate human sacrifice with massacres and genocides in time of war and conquest, as if the one were simply an iteration of the other. Since we think we understand war, we assume we understand human sacrifice. However, this is as problematic as suggesting that since we understand why a man kills his wife's lover in a fit of jealous rage that we also understand the serial killer, since both are guilty of murder. As we know, nothing could be further from the truth.

The phenomenon of blood sacrifice has been described in the modern era as a savage custom of foreign races. The Romans put down the custom of human sacrifice in Britain during the time of the Caesars. The Spanish stopped the practice in Mexico during the time of the Conquest. Human sacrifice was not something that "civilized" people did, although both the Romans and the Spaniards had no problem with invasion, subjugation, and slaughter themselves. They just did not associate their practices with religious ritual.

In the twentieth century, the theme of human sacrifice and savage tribes living at the outskirts of society titillated European and American audiences. These were associated with pagan gods and arcane, thrilling rituals taking place in jungles, swamps, and other remote—i.e., non-urban—locales. Murderous cults in India were given prominence, and fiction writers such as Rudyard Kipling and Talbot Mundy—and officials of the British Raj, like William Sleeman—would make the Thugee famous. The Thugee (from whom we get our word "thug") were both Hindu and Muslim devotees of the goddess Kali, a fearsome deity to some, whose victims had to be killed by strangulation and their corpses later mutilated so that every drop spilled belonged to the goddess. Who can forget the scene in the 1939 film version of Kipling's *Gunga Din* of the Thugees shouting "Kill! Kill for the love of Kali!"? That the "cult" of Thugee was nowhere near as monolithic or organized as the British—who were interested in evangelizing India in the name of Christianity at the time—suspected or declared is now the subject of some academic controversy.

Similar reports—largely imaginary—made their way into the American press regarding Haiti and the religion of *voudon*, or voodoo, as it is popularly known, partly as a justification or pretext for the US occupation of that country in 1914, which lasted until 1934. Gothic horror writer H. P. Lovecraft included a scene of Louisiana-based voodoo worshippers in his story "The Call of Cthulhu"—published in 1928—and associated them with the cult bent on bringing the alien gods back to Earth. Their medium for contact with these gods was the high priest Cthulhu, "dead but dreaming."

One of the strangest of the old religions one can study is that of the Aztecs. On the one hand, their religious philosophy and texts are as hauntingly beautiful and as sophisticated as one could wish. For the Aztecs, poetry was the only "truth" one could expect in this mundane existence.[2] On the other hand, the ritual for which they are best known—human sacrifice—was practiced on such an enormous scale that one could say that no other religion in recorded history was as savage or brutal in its expression.

How could a culture that was so aware of the human condition and the poignancy not to say precariousness of human life be so obsessed with the shedding of human blood, especially in the public arena with all the pomp and ceremony of which they were capable?

▼   ▼   ▼

As in many other cultures, the Aztecs believed that humans were created by the gods as workers and as playthings, as "toys."[3] They also believed that human beings contained the blood of the gods, blood that was mixed with base material in a clay vessel. This is similar to the Sumerian concept.

They also built pyramids, and Aztec theology is based on the idea that the gods pre-existed humans and were organized in families of dualities, of "divine couples," as was the Egyptian pantheon.

It is important to understand that the concept of a divine creation of humans is an almost universal idea. This is not something that we should expect as a normal human understanding of the world or of their origins; it is not something that is self-evident, not even in a purely religious or spiritual context. Why do humans conceive of their own creation as happening because of a god or gods—essentially, as the Gnostics understood it, alien beings—deciding to extend themselves this way? It is one of the myths that elicits scorn from scientists and atheists who insist that human beings evolved over millions of years from single-celled amoeba and that there was no other extraterrestrial or divine cause, and indeed there is no hard evidence to suggest otherwise. Yet, it is a myth that has surprising reach across geographical locales as remote from each other in space and time as ancient Sumer and Aztec Mexico. Is something hard-wired in human consciousness that compels us to think of our origins in this manner? And what of the connection between the gods and blood sacrifice, something we find in every world religion with the possible exception of Buddhism,[4] and which the Aztecs carried to its logical extreme?

Aztec history and culture may give us some answers.

▼   ▼   ▼

The Aztecs were not the earliest civilization in Mexico. That honor probably goes to the Olmecs, the culture that gave us those enormous heads carved out of stone. The Olmec culture began sometime in the mid-second millennium BCE, and thus flourished during the time of the Egyptian New Kingdom dynasties, including the reign of Akhenaten and his son Tutankhamun. Like the Egyptians, the Olmec built pyramids—such as the Great Pyramid at La Venta in the state of Tabasco, which still stands more than one hundred feet above the earth—and in at least one instance (that of the 2,700-year-old pyramid recently discovered in Chiapas, and associated with the Zoque culture, which was either a forerunner of the Olmec culture or an expression of it), the pyramid was used as a royal tomb, with the burial taking place in the temple at the top of the structure.

Intermediary between the Olmec and the Aztec were the Toltec people. This is yet another controversial aspect of the history of Mexico, as some scholars are of the opinion that the Toltec per se were not a separate culture in fact, but a legendary ancestor of the Aztec from a time before they were forced to leave their sacred homeland. The Aztec used the word *toltecatl* to refer to culture itself, to the artisan and the writer/painter, and some theories suggest that the Toltecs were related to the Mayas as suggested by similarities in architecture between the two cultures, the former with their capital in Tula and the latter with their famous center at Chichén Itzá. The histories of the Toltec—like those of the Aztecs themselves—are a mixture of fact and religious symbolism. Extricating the one from the other has been a source of despair among Mesoamericanists for quite some time.

The origins of the people we call Aztec are as mysterious as those of other ancient peoples. The word *aztec* refers to the legendary ancestral homeland of the Mexica people, Aztlan. It is believed that the Mexica migrated to the area around what is today Mexico City around 1054–1065 CE from a region in northwest Mexico (or even from the southwestern United States) at the time of the Crab Nebula supernova (SN1064). Some historians believe that the explosion of the supernova was the cause of the migration, perhaps seen as an omen that convinced the people to move their entire society to the east.

The people who became known as the Aztecs—due to a misunderstanding of the Nahual language by the invading Spaniards—were comprised of several ethnicities who all spoke Nahual and who had some cultural elements in common. They claimed their origin in a place called Aztlan, which was an island in the middle of a body of water, probably a lake, called in some codices "the Lake of the Moon." In the middle of the island was a pyramid in the Mesoamerican style, at the top of which was a tree. This association of an island with a hill or mountain in the center—represented by the pyramid—and a tree in the center of that is a motif we find in other parts of the world representing a mythical kingdom arranged around an *axis mundi*.

They left Aztlan and migrated to a place known as known as Chicomoztoc, "the place of the seven caves." Each of these seven caves was the home of one of the seven ethnic groups that made up the Aztecs, of which the Mexica was one. All were speakers of Nahual. The word Aztlan itself is of uncertain etymology—some suggest it means "place of whiteness," while others say it means "place of herons" or even "place where tools are made"—but the ultimate destination of the Aztecs was the area known as Tenochtitlán (in present-day Mexico City).

The idea that the Chicomoztoc was a "place of seven caves" has resonance with the religious symbolism of other cultures that emphasizes the importance of the number seven as a reference to immortality and to a heavenly palace where the gods reside (possibly associated with the seven stars of the Big Dipper, which revolve around the Pole Star). In the various codices that have come down to us from the time of the Spanish conquest, Chicomoztoc is unique among the places visited by the Aztecs in their migrations, for it is always depicted as being underground. There is thus an underworld association with this place of seven caves, which may also have an astronomical reference.

The Mexica were forbidden by their god—Huitzilopochtli—to call themselves Azteca. The reason is obscure, but may have been due to a desire to conceal the true nature of their origins and the location of their sacred homeland. Instead they were forced to call themselves Mexica and to leave Aztlan forever. This they did specifically on January 4, 1065 CE, according to one interpretation of the Aztec calendar; or the year 1168 according to an alternate reading. Still other codices have the Aztecs leaving Aztlan much earlier, in the eighth century.

Numerous interpretations of the origin story of the Aztecs have been given by various scholars. Some have relied on a historical interpretation, others on an astronomical or religio-spiritual point of view. The truth is probably a combination of all of these. The astronomical event—the sighting of a bright supernova explosion in the heavens—might have been the sign that the Aztecs should leave Aztlan but the desire to find another territory for their culture may have been building for some time. What is certain is that the idea of human sacrifice on a massive scale, for which the term "biblical" would be not only inappropriate but wholly inadequate, was a central fact of Aztec life, and the inspiration for that practice is unknown. What is understood is that this practice was intimately linked to the idea that the gods required constant infusions of human blood in order to survive. The god that seemed to represent this need most strongly was the primary deity of the Aztecs, the aforementioned Huitzilopochtli.

The name of this god resists interpretation. It would seem to be a combination of two ideas, that of "left" or "south" and "hummingbird." Hummingbird seems an unlikely designation for a fierce warrior god; yet, it was said that those warriors who died in battle became hummingbirds and joined Huitzilopochtli in the afterlife. Huitzilopochtli was the god of war and of the Sun. In one of the earliest accounts of the migration of the Aztecs from Aztlan—the *Codex Boturini*—this god is shown as a bodiless head on an altar, in a cave beneath a mountain, speaking to the people.

As mentioned above, in the center of the island called Aztlan there is what appears to be a classic Aztec-style pyramid, and it is surrounded by six other structures, understood as hieroglyphs for "home" or "community." Below the pyramid is a man and a woman. The man is unidentified, but the woman appears to be Chimalma, the mother of Quetzalcoatl, the "Plumed Serpent" god of the Aztecs and known as Kukulkan by the Maya. Her husband was Mixcoatl, "Cloud Serpent," the spiritual force associated with the Milky Way.

There has been much debate over where Aztlan is located and if it ever really existed at all as a geographic site. According to the Codex Boturini, it was located in the northern part of Mexico (or perhaps as far north as California) and the Codex is quite specific as to how long it took for the wandering Aztec tribes to reach Chapultepec, the land in the center of what is now Mexico City. That specificity would seem to rule out the more romantic notions that Aztlan was actually Atlantis, or that the Aztecs arrived in Mexico from the Philippines (as one author has suggested).[5]

First, a figure that appears to be an Aztec priest or ritual specialist leaves Aztlan and arrives by canoe at Cohualcan, the "Place of the Ancestors." It is within a cave in this mountain that the head of the god Huitzilopochtli is seen suspended over an altar, emerging from a hummingbird mask. Cohualcan was also considered the birthplace of Huitzilopochtli himself. It would seem that the priest received the instructions to leave Aztlan from the god at this time.[6]

We thus have several deities represented in this initial panel of the Codex Boturini, beginning with Chimalma who is the mother of Quetzalcoatl by her husband Mixcoatl, and finally Huitzilopochtli who is a major god of the Aztecs. It may seem strange to modern, western eyes to conceive of a god who is more important in a religious setting than the older gods that preceded him, but in practice one could say that Jesus Christ is more important—certainly more popular and the focus of much more attention—to a Christian than God the Father, who seems more remote and "Old Testament." Just as in the study of the Abrahamic religions, however, it is essential for us to focus on these "origin gods" in order to understand how these spiritual traditions developed in the first place. Origin gods present us with a collection of memories—however dim—of the first contact between human beings and events or persons considered anomalous or inexplicable except in the terminology of myth and legend. They establish for us an initial pattern or template upon which all later legends derive.

In the case of the Aztecs, their origin stories involve an ancestral homeland in the remote past and ancestral deities or spiritual forces that are linked to the stars. They also practiced human sacrifice on an unprecedented scale with a complex existential argument supporting it involving human energy and vitality that properly belongs—in Aztec belief—to the Sun God (for instance) and which should be returned to him via the bloody sacrifice of human beings (who need not be followers, or believers, or members of the Sun God's cultus in any way; their only qualification is to be human and acceptable according to the requirements of the particular deity being addressed).

Chimalma, the only deity we are certain was connected with Aztlan, was seized from her home by Mixcoatl (according to the legend) who shot arrows between her legs, and then impregnated her. This was later modified or expanded to include her becoming pregnant with Quetzalcoatl by consuming a piece of jade.

The stories of how humans and gods were conceived share something in common across many ancient cultures. Conception rarely occurs in the way with which humans are familiar. There is always an understanding that conception involving the divine (whether of gods or of humans) was not the result of normal sexual intercourse. Just as Chimalma became pregnant with Quetzalcoatl by eating jade, so did the wife of the Indian god Shiva become pregnant by receiving Shiva's seed either in her mouth or her hand. Even Mary, the mother of Jesus, became pregnant while she was still a virgin and her conception was announced to her by an angel; etc.

To the Aztecs, jade had symbolic significance. It was used as a euphemism for blood as well as for heart. In children, it was believed that their blood and their hearts were still green and often the word jade is used to refer to their unripe—and therefore pure—status. The divine energy was very much alive and strong in children; as people became older, the divine energy became snarled and twisted in their bodies, entrapping the god within. The only way to release that imprisoned deity was through sacrifice, and the best mode of sacrifice was one in which the spirit was wrenched free of the body: the limbs severed, the head severed, the heart removed from the chest. Like the Gnostics, the Aztecs believed that matter was the prison of spirit; unlike the Gnostics, the Aztecs believed that the best way to free the spirit was to destroy the material basis in which it was trapped. One could say they took the Gnostic attitude to its logical conclusion.

The sacrifice had to be violent and bloody. There was no gentle poisoning, no mere suffocation (there were a very few cases where victims were squeezed to death or smashed against a "God stone," for instance). There had to be blood, for blood was the spirit. Blood would flow directly to the gods, like serpents writhing from the necks of the victims, and in fact Aztec illustrations depict precisely this image of blood serpents flowing from the necks of decapitated victims. Like the goddess of India, Chinnamasta, who is depicted as a headless woman, holding a knife in her right hand and her severed skull in her left—exactly the placement shown in the Aztec drawings—blood flows out from her neck in separate streams into the mouths or the cups of her followers.

Blood is spirit; blood is life. More importantly, blood is of divine origin, not human. It does not belong to human beings but partakes of the divine—the alien—essence. That is the understanding among an amazing variety of ancient cultures.

Biologically speaking, we know this is not true. Human beings cannot exist without blood. They can exist without their limbs, with the loss of their senses, etc., but not without their blood or the hearts that pump that blood. Yet, even in the present day, we speak of family members as being consanguineous, i.e., sharing the same blood. We speak of bloodlines as if they are genetic lines, and that is where this particular idea becomes more relevant to our study.

Many people are aware of the motif of the voodoo doll, for instance. This artifact is used to heal or harm—or otherwise control or manipulate—another human being through action at a distance. It is fashioned as closely as possible to the desired target: same gender, same general appearance, etc. Then, and perhaps most importantly, something of that target is incorporated into the manufacture of the doll: usually the target's hair or blood. This establishes a link between the inanimate doll and the person intended.

This is a very old concept, related to what anthropologists call "sympathetic magic." There is a "sympathy" between one object and another, usually suggested by similarity in appearance. If something looks like something else, they are in a sense related; thus action on one will produce a corresponding action on the other. In the case of the voodoo doll, stabbing the voodoo doll is supposed to cause the person represented to fall ill and die.

While there have been many explanations for the reasoning behind this—including the intense focus on the doll as being somehow responsible for any effects that occur—one that has been ignored is the relevance of the blood and hair.

While using the target's bodily fluids and essences may serve to help focus the attention even stronger—from the point of view of the ritual specialist—there is another aspect to this ritual that would have been unsuspected until the twentieth century, and that is that every person's DNA is unique. If there is DNA in the hair (supposing the follicle is intact) and in the blood, what the practitioner has in their possession is the equivalent of that target's entire genetic code. One could, in theory anyway, construct a clone of that person if provided with just enough DNA to allow replication in the laboratory and subsequent in vitro fertilization of a viable ovum.

There is, however, another aspect of this genetic approach that has only recently been proposed, and that is the newly formulated idea of *DNA consciousness*.

In a series of peer-reviewed papers and presentations, a New York researcher has proposed that DNA not only provides the mechanisms for human consciousness by building our nervous systems, brain structures and sensoria, etc., but that DNA itself has a level of consciousness that enables it to carry out its functions.

John K. Grandy's theory[7] was extended to include all forms of matter, down to the smallest subatomic particles. Quantum states were described as revealing levels of consciousness, as well. This is due to his newly articulated definition of consciousness as the result of interactions between objects, and their interaction with the environment (including energy and forces).

The motivating idea behind DNA consciousness is the impossibility of accepting that DNA acts randomly. Each cell of the human body, for instance, contains DNA identical to every other cell but the degree of specialization of each of those cells—depending on genes within the DNA molecule—indicates a higher-level process whose properties are so far unknown. Grandy's theory is that the *DNA itself* is conscious. That doesn't mean that the DNA molecule ponders the meaning of life, for instance; on the contrary, DNA is life itself. Subtract DNA from the Earth and all that is left are oceans and mountains. No plants, no insects, no animals, no single-celled amoeba. And no people.

The details of Grandy's research are too complex to detail here. Basically, he posits several levels or layers of consciousness, beginning with subatomic particles as one level, (the quantum or primordial consciousness), atoms as the next (atomistic consciousness), then molecular or DNA consciousness: the type of organization of complex interactions represented by DNA. After that there is human consciousness itself (and the possibility that there may be a higher form of consciousness above the human, which is, of course, what mystics have been saying for millennia).

▼   ▼   ▼

Is it possible that pre-scientific peoples understood this quality of the blood as a kind of identification mark, a measure of uniqueness, as well as a form of consciousness related to the divine? It is clear that, for the purposes of human sacrifice in Mexico, the victims had to be carefully chosen. If all that was required was human blood, then any human blood should have been acceptable; however, this was not the case. Qualities of the individual—heroism, strength, intelligence, power, royalty, etc.—were believed to be inherent in the blood. The blood was the carrier of these qualities, and it was these qualities that the gods desired in sacrifice.

Victims were chosen on the basis of what we would call genetic mutations. A child born with a double cow-lick was especially prized. Children of particular beauty were selected for sacrifice. The more perfect or special the child, the more likely it would be murdered as an offering to the gods.

For a comparison to this in other cultures, we can glance at the drawings of demons in European literature including the grimoires, or in the paintings of Hieronymus Bosch, where they are uniformly represented as deformed humans: just enough of human physiology to be recognized as such, but with exaggerated limbs, twisted torsos, and especially misshapen heads and features. It is as if there is an unconscious recognition that physical deformity is equivalent to dead-ends on the evolutionary path, genetic mistakes that can still return to haunt us. Literally.

In the works of the witch-hunters, a great deal of attention is paid to the work of the succubus and the incubus: two versions of the same demonic force which seem to represent some of the same experiences that we find in abductee literature. In the *Malleus Maleficarum* (1486) of the inquisitors Heinrich Kramer and James Sprenger we find detailed descriptions of the way demons steal the semen from sleeping men and transport it to unsuspecting women to impregnate them. The demons themselves are sterile but desire to produce offspring in this manner. The idea that a woman would become pregnant this way is somewhat blasphemously reminiscent of the way the Gospels relate Mary's conception of Jesus.

The salient characteristic of these cases is that physical ugliness and deformity—genetic conditions—are considered evil and demonic, while the genetic conditions that produce beauty are divine. This seems obvious or self-evident to us today, and that may be part of the problem for it would be considered the height of chauvinism and insensitivity to declare those whom society deems unattractive to be spiritually corrupt or the offspring of demons. Yet, beauty and ugliness are metaphors for something deeper, an operation taking place on the genetic level that is beyond our conscious understanding but which erupts in times of great spiritual tension, such as Aztec Mexico or medieval Europe during the Inquisition: cultures existing simultaneously on different hemispheres (until the Inquisition came to Mexico from Spain in the sixteenth century). This same concept would erupt again during the Third Reich and the elevation of an idea of Aryan beauty above that of the other races, races—particularly the Jews, considered as a race apart—which were pictured very nearly the same way demons were depicted in medieval times.

In cases where the sacrificial victim was a stand-in for a deity, the victim would be treated as a god. He or she would be dressed in fine clothes, fed sumptuous meals, adorned with the mask and other accoutrements of the deity, and finally brought to the sacrificial altar with all the pomp that the god itself would expect. Did this—in the minds of the Aztecs—create a situation in which the human blood of the victim was infused with the qualities of the god?

Was this an instinctual understanding of what Grandy calls "DNA consciousness," but in reverse: consciousness affecting the DNA? Was this the template for the idea of the voodoo doll: in this case an image of a god instead of a person? (In Mexico after the Spanish conquest and the suppression of their religion, Mexican devotees would worship the old gods in secret. They would make dolls representing the gods, items they could easily conceal.)

In the complex myth system of the Aztecs, the blood of a slain stand-in would be especially valuable to the god being represented (otherwise why go to all the trouble?). It was not the blood of the victim before he or she was "consecrated" to the divine service that was valuable, but the blood obtained only after the victim had been treated as the god. This manipulation of the consciousness of the victim was reproduced in the victim's cells, in the victim's physiology. In the blood.

To be clear: it was not merely death alone that was desired. The appearance of blood was necessary. One's divine essence was not released to return to its divine origins upon death, but upon dismemberment, decapitation, disembowelment, and the extraction of the heart itself.

It is a measure of the sophistication of this ritual that no one considered the sacrifice of the god stand-in as the murder of the god itself. The Aztecs were not killing their gods; they were propitiating them. They had to elevate the human blood to the status of divine blood, and this could only be done by changing the status of the victim from human to divine.

Is there some sort of evolutionary *raison d'être* behind this hideous practice?

▼   ▼   ▼

As we will see in more detail in Book Three, the idea of panspermia—although suggested in the twentieth century by the co-discoverer of the DNA molecule, Francis Crick—has roots in ancient ideas about human origins. As we saw in the chapter on Sumer, their creation epic involves human beings created from the blood of a slain goddess, Tiamat. In the Christian ritual of the Mass, wine is ceremonially changed into the blood of Jesus *and is consumed.* The sharing of blood is the sharing of a spiritual essence. The sacrifice of a creature and the shedding of its blood is a returning of that essence to its source. In the case of the Aztecs—as well as the Olmecs and the Maya—the source is divine: the gods, who belong to a different universe, a different category of existence (called by the Aztecs the "Palace of the Fleshless"), partake in ours invisibly, spiritually, yet inextricably. Aztec devotees upon nearing a shrine to one of their gods on a roadside, for instance, would cut themselves and drop a little of their blood onto the stones as an offering. What it means to be a human being is intimately involved with divinity and divine origins, yet not as identical beings but created for purposes beyond our comprehension. It is only by participating in the creative process ourselves that we become gradually aware of the reason for our existence.

This identification of blood with spirituality became an obsession among the Aztecs. Although evidence shows that the Olmecs and the Maya also practiced human sacrifice and for many of the same reasons, the Aztecs carried this ritual to such an extent that it was the focal point of their culture and the basis for an entire ritual vocabulary. The Aztec calendar was consulted for the best days for specific types of sacrifices and for specific victims: male, female, adults, children. One might think that Aztec rulers and the elite were spared the ordeal of being slaughtered on top of the altars already slick with the blood of the innocents, but not so. No one was immune; anyone could find themselves selected to be sacrificed at some point, or to die in battle. To die of sickness or old age was considered shameful. Magicians, for instance, were to be sacrificed when they reached the age of fifty. Captives were taken in battle and sacrificed; at other times or for different rituals victims simply were purchased from the towns and villages under Aztec control.

As we saw, some victims were dressed in robes and masks that were specific to a certain deity, and were treated as royalty before their execution. It was also customary in many cases to make the intended victims drink a potent liquor or drugs shortly before the sacrifice so they would not panic or try to break free (which would be considered a bad omen).

For some festivals, hundreds of victims were sacrificed. One account has as many as 80,400 individuals sacrificed during one bloody celebration—the consecration of the

Templo Mayor in what is now Mexico City—a number that is certainly an exaggeration or an error in calculation but which nonetheless gives some idea of the scale of the ritual murder we are studying. In some cases, the bodies were broken on rocks or knives slashed the skin. An important Aztec deity was Xipe Totec, the "Flayed One," whose statues show him wearing strips of human skin that had been flayed from the victim's body.

In other words, the Aztecs deified horror.

The problem with all of this is the fact that the Aztec fixation on blood, sacrifice, dismemberment, and decapitation—and the association of these elements with spirituality and divinity—is simply one of degree, not of kind. Some of the deepest practices of shamanism as well as forms of Indian and Tibetan Buddhism involve these same elements, albeit not in reality but as facets of meditative practices and initiatory experiences. The Tibetan *dcod* practice, for instance, involves all of these violent images as a way of destroying the ego through intensive visualization. According to Mircea Eliade, Siberian shamans undergo these same experiences during initiatory periods spent alone in the forest; their inner organs are removed, washed, and re-inserted in their bodies, as an example.

In India, the goddess Kali—mentioned above in connection with the Thugee—had a similarly frightening image: her naked body, pendulous breasts, lolling tongue, and necklace of skulls is another deification of horror. She also is shown holding a severed head in one hand and a sword or knife in another. Her temple in Kolkata—the

Kalighat—is at times awash in the blood of sacrificed animals. She is often depicted as standing on the corpse of Shiva in a carnal ground, thus emphasizing her aspect of movement and energy (*shakti*) to Shiva's stationary and contemplative nature: the two being necessary for creation.

In Christianity, the most famous symbol is that of the cross. On this instrument of torture and death the body of Christ was nailed. It is a central image in virtually all Catholic churches around the world: an image of torture and murder, and blood.

These horrific scenes are experienced in virtual fashion in much of the world, but in the Aztec culture they were experienced in reality: either as the victim, undergoing the increasingly imaginative sacrificial process, or as the priest committing the sacrifice, or as the observer: a member of the Aztec congregation witnessing these events many times during the course of a lifetime. The gods themselves were dressed in human skin, or adorned with human skulls and body parts, and smeared with the blood of the altars. Imagine the serial killer Ed Gein—the inspiration for *Psycho*—being worshipped as a god.

Hannibal Lecter as Jehovah.

An essential element of all of these practices—from the sanguineous rites of the Aztecs to the *dcod* meditations of the Tibetans—is the explicit acknowledgment that human beings are inferior to their invisible, unseen gods and that this inferiority is acted out in rituals that emphasize human inadequacy. The self-flagellation—both actual and virtual—of these widely varied forms of spirituality suggests a degree of self-loathing inherent in the human condition that is nothing short of astonishing.

There is only one salvageable—even salvific—aspect of human existence, and that is the blood. The blood is the one human product that the gods desire; it represents—for the Aztecs—the quality of *istli*, or the energy or heat of the sun. Ritual murder freed the *istli* from the prison of flesh and allowed it to return to the gods. Some Aztec iconography depicts the blood of victims flying upward to the heavens, as if returning to its home in the Palace of the Fleshless in a kind of vampiric Ascension.

How have human beings made this connection between themselves and the divine? Indeed, it is virtually the *only* way in which humans interact with these invisible beings. Gods everywhere demand sacrifice. Cain knew this when he slew Abel; the only mistake he made was in choosing his own brother as the sacrifice out of jealousy: that meant that the sacrifice was divided between an offering to God and a satisfaction for himself, and God does not like to share.

Abraham was expected to sacrifice his son, Isaac.

Throughout the Old Testament the Semitic tribes were constantly in the throes of massacre, slaughter and sacrifice. In fact, there is even an enigmatic reference to child sacrifice in Exodus 22:29 where God tells Moses:

> Do not hold back offerings from your granaries or
> your vats.
> You must give me the firstborn of your sons.

And in Ezekiel, 16:20-21, God complains to Ezekiel of the iniquities of the Jews which led to their captivity in Babylon:

> And you took your sons and daughters whom you bore to me and sacrificed them as food to the idols . . . You slaughtered my children and sacrificed them to idols.

Ezekiel's visions took place in the sixth century BCE (about the time of the fall of the Olmec empire in Mexico and the beginning of the Maya and Zapotec cultures). He is recording a charge that child sacrifice (as well as temple prostitution and other religious crimes) were taking place in Jerusalem at the time of the conquest of that city by the Babylonians.

This is mentioned to emphasize the fact that human sacrifice was practiced among the people who gave us the Abrahamic religions as well as among those in Mexico, India, Britain, Africa, and elsewhere. It is a *human* trait, a compulsion that historians have struggled to explain. In order to make any kind of sense of it *from a scientific point of view* one would have to take a position outside the cultures being investigated which, in this case, means a position outside human culture itself. It is not reasonable to expect that a person who is the descendant of a culture that practiced human sacrifice would be able to investigate it without preconceived ideas of it coloring the analysis. Since we would be hard put to identify a culture or race on the planet that did not practice human sacrifice to some degree, we would require the assistance of an anthropologist from outside our system. This point of view says, in effect, that human beings are incapable of analyzing themselves to any meaningful degree and this may be true. We seem to agree that murder is wrong; that the killing of innocents is abominable; that peace is the desired state of being; and yet we are still incapable of achieving any degree of control over our baser compulsions. This would seem to indicate either a failure of self-analysis or a vulnerability that has been programmed into our genes. As fanciful as that sounds, let us look at the evidence that all of this bloody sacrifice suggests.

▼   ▼   ▼

The history of the world's civilizations is a moveable feast. While pride of place is given to Mesopotamia as being the "cradle of civilization" and the site of the earliest recorded culture—that of the Sumerians—we now know that human society thrived all over the world at the same time as the rise of the Sumerian city-states. Recent archaeological discoveries in North and South America are proof that cultures were in existence on those continents as long ago as the twelfth century, BCE. The Winnemucca petroglyphs in Nevada have been dated to about that time, with rock carvings depicting leaves, a wheel or sun disk, and other geometric figures. There was no writing left behind, no cuneiform tablets or papyrus scrolls, but the level of sophistication required to conceive, design and execute those carvings indicates a higher level of civilization than was previously imagined for that time in North America.

The same is true of South America, where evidence of human habitation has been found to have occurred more than twenty-five thousand years ago, predating the supposed migration of people from Siberia across the Bering Straits and down the coast of North America in 13,000 BCE. Paleontologists have discovered evidence of human habitation in Chile, Brazil and Uruguay that show these areas were populated from 15,000 to 30,000 years ago. These discoveries not only challenge the prevailing Clovis theory (that humans of Asian descent reached the Americas from the Bering Strait 13,000 years ago) but they also challenge theories about the racial composition of these early inhabitants with DNA and other metrics suggesting elements of non-Asian ancestry. There is even evidence that some DNA identical to those found among the "Beringians"—i.e., those who supposedly lived in Beringia, the area now flooded under the Bering Straits, who formed the ethnic group that populated the Americas during the so-called Clovis period—was discovered among Icelanders. This is fueling speculation that a race living in the Americas traveled to Iceland around the time of the Norse expeditions to what is now Greenland, Canada, and the northeastern United States. Alternately, it could be evidence that people from Europe visited North America a thousand years before Columbus, or longer.

What this means is that there is no way definitively to identify an "origin point" for human civilization. Add to this the mysteries surrounding Mohenjo Daro and Gobekli Tepe and you have enough ambiguity for several lifetimes of research and frustration.

Yet, another source of archaeological data is the human genome. Nothing is lost in the genetic database. And the fossil record also shows us how to trace human ancestry back millions of years to the earliest hominids in Africa. What the genome and the fossil do not show us—at least, not at the present time with our available technology—is the influence of culture. And they certainly do not show us the possibility of contact with beings or entities— tool-making, communicating, mobile entities—that were not human, that is, neither Neanderthal nor Cro-Magnon (which separated from each other around 300,000 years ago or longer and developed along independent lines).

Research concerning the origins of consciousness indicate a point five hundred million years ago, during the Cambrian Period; but that depends on how you define consciousness which, as we have seen, is controversial. In this recent study,[8] it was proposed that the genetic basis for consciousness would have to include the development of a brain and a nervous system and that this took place among the vertebrates during the Cambrian with the development of an optical system as well as a rudimentary tripartite brain. This, of course, is not identical to what we as humans consider consciousness: the creature being credited as the earliest conscious vertebrate is the lamprey.

Ancient alien theorists like to speculate that aliens came down from the sky at some point along the developmental lines of modern humans—i.e., descendants of the Cro-Magnon if not Cro-Magnon themselves—and influenced human behavior by teaching technologies such as metallurgy, agriculture, astronomy, etc., and then disappeared again. They point to various archaeological sites as evidence of either direct alien contact or knowledge of alien technologies (since lost, or hidden).

They sometimes expand this theory to include the Zecharia Sitchin model, which is that aliens created the human race as a slave race, designed to serve them in mining and other endeavors. The Sumerian creation epic, as we have seen, could be interpreted that way *to an extent*, but that means it would have taken place long after the seeding of our planet with DNA hundreds of millions of years ago and the development of rudimentary forms of consciousness in vertebrates about 500 million years ago. (Sitchin's equation of the biblical Nephilim with the Sumerian *annunaki* is tenuous, at best, as has been discussed in our chapter on Sumer.) If we accept as a theory that there was alien intervention in the evolution of the human race at some point along its genetic timeline from the first vertebrates to the first primates, to the Cro-Magnons and Neanderthals, and from there to the civilization of Sumer, then there should be evidence in the genetic code.

As it happens, there may be. As we will discuss at more length in Book Three of this series, two researchers in astrophysics, astrobiology, cosmology and mathematics in Kazakhstan claim they have identified evidence in the genetic code which points to an off-planet origin. Their research[9] was first published in 2013 and they have followed up since then with more articles on the subject. Their theory is based on the mathematics of the genetic structure which, they say, incorporates non-random sequences and which in particular makes use of a symbolic notation for "zero": the stop codon.

The concept of zero as a place-holder in arithmetical systems is a relatively recent one in human history. The Egyptians conceived of zero and used it in their architectural design as a base line, represented—oddly enough—by the symbol of a heart and trachea. India is usually credited with giving the world the concept of zero as a number, but it was also used much earlier in Mexico, among the Olmecs, the Maya and the Aztecs for use in calculating the Long Count of their calendar system. Thus, the researchers say, the use of a zero symbol in the genetic coding system indicates something non-random and deliberate, the presence of a "message."

From the perspective of ancient alien theorists this would suggest that our genetic code (or some part of it) is an inheritance from this putative source, and that we carry the knowledge of this "alien" identity around with us as an inextricable part of our very being. In other words, we are "branded" in a sense with this indelible mark of servitude, a kind of "mark of the Beast." Their take-away from this is that we are useful only insofar as we serve the ones who created us. Otherwise, we are in danger of being discarded. Or worse.

Like the CEOs of a major industrial empire, then, the alien forces responsible for our existence would look askance at any attempt by their "employees" to form a "union." The last thing the creators of any product need is for their producers to form alliances that threaten the status quo, that enable the serfs to build their own castles. Also, there is no need for a product that will last forever. "Planned obsolescence" is built into many products, and why should human beings be any different? As we know, we are not. Our obsolescence is planned. We may pull a few more years out of the coding every century or so, but our fate has been pre-determined. Like the Bard once wrote— in a slightly different context—"The fault, dear Brutus, is not in our stars, but in ourselves."

Why not entertain the possibility that the truth is somewhere in between?

There is a growing body of evidence to show that our genetic code came from elsewhere: either from an asteroid strike or some other extraterrestrial event. This is the position of Francis Crick, one of the discoverers of DNA. That the genes might have been "seeded" by another race of beings is certainly possible, even likely. That does not remove the possibility that the planet was visited in ancient times, however.

What would have been the reaction of those human beings upon confronting their creator race for the first time? (We are proposing that this visit would have been by the creators and not some other alien race, although the chances are good that we would share some genetic material in common, perhaps from the same source.) Would there have been a sense of familiarity, mixed with an opposing sense of the Other and the Alien? It is a given that this race would be more technologically advanced, and for that reason would appear to be all-powerful, all-knowing.

Our evolution would have taken place in an environment different in some respects from the "mother" environment, which would have resulted in significant physiological differences. It's entirely possible that human mortality itself is the result of genetic adaptation required by the environment on Earth: a mutation in the gene sequence that leads to the ageing process, perhaps. Or, alternatively, mortality was part of the original coding.

In the eyes of these putative creators, then, are we a successful experiment or a failed one?

Perhaps we are both.

▼   ▼   ▼

The example of blood sacrifice—practiced on a global scale—seems to be an acknowledgment of our indebtedness to an absent creator. This creator desires our blood; or at least that is how we understand it. Our blood is not ours; it belongs to some other entity. Our blood is the divine essence—some specific encoding of our genes—trapped within the weak and mortal human body, imprisoned and waiting to be released, to go "back home."

It is proposed that some of the deeper mystical practices of the shamans and the Tibetan ritual specialists—those involving the interiorization of the bloody sacrifices through meditation and other practices—might represent attempts to understand this need of the "gods" by experiencing the events psychologically and emotionally. The *dcod* meditation (as an example) might be a way to undergo (virtually) the dismemberments, decapitations, and violent bloodlettings in order to catch a glimpse of some other truth, some indication as to the meaning or purpose behind this self-destructive requirement.

Somehow we understand that these gods want us to destroy ourselves. When a civilization becomes stronger than its neighbors—when its empire stretches across vast expanses of land and sea—then the self-immolations begin. When the Aztec empire reached its zenith in the centuries before the Spanish conquest, having subdued all their enemies, they still practiced human sacrifice on a massive scale.

And when human sacrifice is not enough, there is always war.

▼   ▼   ▼

The advance of *Homo faber*—the toolmaking human—meant that war could be conducted in such a way and with such weapons that a small nation could subdue countries much larger than itself, with much greater populations and land masses. Transportation—by land, sea and air—was a great enabler. England managed to colonize half the world with its navy, and England is a relatively small island. Consider the disparity between the size of England and the size of its largest former colony, India.

When the Spanish invaded Mexico, they did it from ships across the sea and with horses on land. The Aztecs had never seen horses before. They represented a terrible engine of destruction.

What the development of the tool accomplished for the human race was the extension of the hand, and by extending the hand beyond where it normally would be effective we also extended the mind. Our reach suddenly increased to the point where we could stand behind a hill and aim an arrow at another human being far away and kill him; or, by using a system of pulleys and wheels, manage to lift enormous stones into place with much less effort—much less physical strength—than any ten humans would possess. It didn't matter if we were short or tall, fat or thin, in good physical shape or not; a tool enabled the mind to accomplish what the body could not. In fact, it even enabled the individual mind to accomplish what its own limited intellectual abilities could not, by relying on the technological superiority of the toolmaker as a substitute for our own individual ignorance or stupidity. An automatic weapon is a marvel of technological know-how, its creation requiring an understanding of engineering, physics, and chemistry as well as of manufacturing and raw material supply. Then it winds up in the hands of a someone with a borderline IQ, and perhaps borderline personality disorder as well, and tragedy ensues as the anemic youth opens fire on a crowd in a movie theater. The scope of that tragedy would have been impossible in earlier times: the same damage would have depended on that person throwing a couple dozen rocks with lethal accuracy unimpeded or perhaps beating a dozen people to death with his bare hands. It's not something our imaginary youth would have considered, much less acted out, but a tool makes it possible and in many ways inevitable. The gun, as they say, is the great equalizer.

Other technological advances, such as siege engines, tanks, airplanes, and bombs, forever changed the nature of war from a struggle between two tribes or clans over resources to one of ideologies from other sides of the world, of trade and economic interests incomprehensible to the other, to the use of terrorism as a tactic to undermine the nerve and the will of an opponent through the use of targeted attacks on non-combatant populations using the same technological tools—the same weapons—as the largest armies of the most powerful nations.

There are always justifications for going to war. Those who lose their lives are usually the same as would be sacrificed in ancient Aztec rituals: young men, and innocent women and children. In other words, the most fertile, the strongest, the most capable of any society. We talk of our soldiers making "the ultimate sacrifice" without realizing how apt that sentiment really is and without admitting who is responsible for the sacrifice. We conduct wars almost blithely, using national security as the rationale even for the invasion of countries on the other side of the world, countries with limited resources and materiel and which themselves pose no military threat; and then we wonder why there are murders taking place daily, hourly, within our own borders, with our own people killing each other. We are powerless to defend ourselves from ourselves. We know there are other possibilities, other ways of looking at the world, but are unable—or perhaps even unwilling—to commit ourselves to the alternatives. Legions of philosophers, generations of scientists, successions of apostles, and the human race is no further advanced than it was at the beginning of recorded history.

Is this intentional? If DNA has consciousness, is this the goal of its brilliantly complex organization?

We are toolmakers. The builders of machines. The past ten thousand years is evidence of that. We are not the makers of consciousness. Instead, we have made consciousness take second place to that of the machines without realizing that the *machines themselves may represent a form of consciousness*. What understanding we have of our own reality—not the reality of chromosomes and chemicals, of photons and gravitons, but the reality of our existence, our purpose—is limited, circumscribed, by a power not of our choosing. As mentioned earlier in this chapter, in order to understand the process of creation we have to take part in it ourselves. Our need for machines may be an expression of a deeper understanding of who we are. The machines we see—the ones we are told *are not really there*—may be the inspiration we need to take the next step in creating ourselves.

We have sensed this, almost from the beginning of our known histories, but have been unable to defeat the genetic program. Yet, as we will see in the next volume, beginning in the twentieth century we made a deliberate and conscious decision to unravel the secrets of the human mind and to penetrate the veil of consciousness even as we made the most powerful tools known to humanity, the machineries not of joy but of mass destruction. We know—on some kind of subconscious, subliminal level—that this can't go on. We know that our time is running out. We know that we have been lied to.

We know, like the doomed Aztecs, that we have been slaves to dead gods.

The Mayan temple at Tikal, Guatemala. Notice the similarity to the Ziggurat of Ur as well as the alchemical image of spiritual integration and physical transmutation from the previous chapters.

A depiction of Aztec human sacrifice on the steps of a temple in Mexico, as discovered in the Codex Magliabechiano. The heart of the victim is seen ascending to the heavens.

The La Venta pyramid in the state of Tabasco, Mexico, which was built by the Olmecs, stands more than 100 feet above the earth. There are pyramids like this throughout the ancient world, from China to the Middle East to the Americas and more are being discovered every year thanks to satellite imagining technology.

Viracocha, the creator god of the Incas. He is the god of sun and of storms, and once caused a "Great Flood" to destroy humanity. He is depicted wearing the sun as a helmet or mask around his head, and holding thunderbolts (like Marduk, and some depictions of Abrasax, the Gnostic deity).

From the Codex Boturini, this image shows the voyage of the ancestors of the Aztecs traveling from the legendary island Aztlan with its stepped-pyramid (ziggurat-like) structure in the center.

A Toltec image of Chicomostoc: their legendary seven-caved ancestral home. The idea of an origin among seven caves or seven hills is common to many ancient peoples and may refer to a celestial origin among any of the seven-starred constellations or asterisms. The Egyptians connected their ancestral home to the seven stars of the Big Dipper, and there were similar ideas among the ancient people of India and China.

LET US NOW RAISE FAMOUS MEN

The boundaries which divide Life from Death are at best shadowy and vague. Who shall say where the one ends, and where the other begins? We know that there are diseases in which occur total cessations of all the apparent functions of vitality, and yet in which these cessations are merely suspensions, properly so called. They are only temporary pauses in the incomprehensible mechanism. A certain period elapses, and some unseen mysterious principle again sets in motion the magic pinions and the wizard wheels. The silver cord was not for ever loosed, nor the golden bowl irreparably broken. But where, meantime, was the soul?

– Edgar Allan Poe, *The Premature Burial*

The mythology had therefore to serve two purposes. It was to give the steps whereby the universe was arranged, leading up to the final triumph of Horus and the coming of the pharaonic monarchy. The other purpose–only gradually understood–was to provide a series of symbols to describe the origin and development of consciousness . . . . When they mythologized they knew what they were doing.[1]

– R.T. Rundle Clark,
Myth and Symbol in Ancient Egypt

DEATH——THE FACT OF DEATH, THE FEAR OF DEATH——IS A central, overriding concern of the world's religions for it is death that inspires the desperate search for meaning in life, for relevance, for God. They say that awareness of death is necessary for appreciating life; it is even more necessary for the establishment and maintenance of religion. This is why Karl Marx could write that religion is the opium of the people: it keeps them docile in the face of state power, certainly, but it also distracts them from the inevitability and total mystery of death and, in many cases, motivates them to risk their lives in service to religion or the state because they are assured of life after death. That, however, is a purely materialist point of view.

If religion has its origins in the experience of alien contact, as we have proposed, then the Phenomenon really has nothing at all to do with death. With all of the records that have been collected on the Phenomenon—especially those of the last seventy years or so—there has been no evidence that alien contact has conferred immortality on anyone. If, as has been suggested, there were alien corpses present at the Roswell crash site (or any other crash site) it would seem to argue against any claim by aliens of immortality, or any belief in alien immortality by experiencers. Aliens die. That alone should make us hit the pause button for a while and wonder if contact has anything to offer us than simply more and better technology, or yet another religion.

There are persistent stories that alien entities have the ability to heal humans from sickness, etc., or even to bring humans back from the dead. That is not the same thing as immortality, however, even if it were true. We have the alien, Klaatu, in the 1951 movie *The Day the Earth Stood Still*, raised from the dead by the robot, Gort, after Klaatu was shot by a soldier, but Klaatu reveals that his condition is only temporary. Even aliens with advanced technology and a stardrive cannot overcome death.

If so, then what's the point of alien contact?

It was Anne Strieber—the wife of alien abductee Whitley Strieber—who noticed a weird correlation between experiences of alien contact and the presence of the dead.[2] Having the dead appear alongside aliens during a contact experience seems to indicate a post-death existence of *some kind*. Is that the same thing as immortality, however?

In the Cargo Cult example given at the very start of this book, the good people of the Stone Age cultures of the South Pacific witnessed the arrival of planes, motorized vehicles, radios, clothing, food and medicine from "above." A lot of good things, impossible things by their standards but nevertheless desirable. However, what they did not see was anything resembling an elixir vitae, a cure for death. Even the godlike beings witnessed by the jungle-dwelling visitors could not promise immortality. In fact, their business was war. And invasion.

It may be that what we seek is not a "cure" for death but a means of maintaining our individual identities after death. We can prolong life; we have been doing that for some time, now, and may continue to do so until life expectancy reaches 100, or 125, or more. The human body does not need to shut down; science has been telling us for some time that certain obsolescence programs in the human genome can be altered, modified, perhaps one day even reversed. Perhaps the aliens who have been experienced are ancient themselves by Earth standards; maybe they routinely reach the age of 300 (as the 1952 film *Red Planet Mars* suggests).[3]

But while reaching such an extraordinary age (without illness or senility) may seem to be a pretty good deal, there will still come the moment—at age 295, say—when we start to refuse to go gently into that good night. There may be even more reason, after such a long life and many more experiences, to hold onto our extensive library of thoughts, dreams, and memories: to maintain our identity and thereby our memories after the body fails us, so all those moments won't be "lost, in time, like tears in rain."[4]

If there is longevity to this extent, and if there is existence of identity after death, then perhaps space travel holds the key that unlocks the elixir of life. Perhaps this is what all those venerable civilizations were trying to tell us with their weird, seemingly pointless, architecture and their complicated funerary rituals. Maybe this is the most important element of that first, pre-literate, pre-historic Cargo Cult contact. Maybe all this desperate focus on God or gods is merely what we have taken away from the Pyramid Texts, the Book of Genesis, the Gospels, the Qur'an, the Bhagavad-Gita, the Lotus Sutra, the Zend Avesta, and does not represent what inspired this desperation in the first place. For what is the cause of most human desperation on this planet?

Fear of death.

▼   ▼   ▼

Ancient Egypt is, for many, the *fons et origo* of Western civilization including most especially spirituality in its many forms. For hundreds of years all that was known about Egyptian religion and culture was derived from statements by Greek historians or from tentative analyses of Egyptian art and architecture such as the pyramids and the writing system known as hieroglyphics. However, hieroglyphics were not translated until long after a great deal of fanciful speculation already had taken place.

In eighteenth-century Europe there were a number of mystics and occultists who rode the then current Egyptian fad by creating secret societies based on presumed "Egyptian" mysteries. The famous Count Cagliostro (1743–1795) was one such initiate who created a form of Freemasonry in 1784 based on his interpretation of Egyptian esotericism. Egyptian hieroglyphics were not translated until the 1820s when Jean-François Champollion was able to use the newly discovered Rosetta Stone, which meant that occultists and esotericists were free to project onto the mysterious writing whatever meaning they liked. (A famous example of this creative imagination was the founder of the Church of Jesus Christ of Latter-Day Saints, Joseph Smith Jr., who claimed he was able to translate "Reformed Egyptian" hieroglyphics in the case of his Book of Mormon. There is no such thing as Reformed Egyptian.) Prior to the decipherment of Egyptian hieroglyphics, however, some notable texts were produced that seemed to concretize the general European tendency to see Egypt as the source of all wisdom, particularly of the non-Christian kind. One of these—mentioned in a previous chapter—was the Emerald Table of Hermes.

Ancient Greece is an important resource for knowledge about Egypt and Egyptian religion, although by the time that the Ptolemaic (Greek) dynasties were in charge in Egypt—the third to first centuries BCE—much of Egypt's culture was in the process of being assimilated into a more cosmopolitan Greek model. Alexander the Great had invaded Egypt in 323 BCE and installed his own people as rulers. It was Alexander who began the construction of the famous Library of Alexandria, which later was destroyed during the invasion of that city by the Romans under Julius Caesar circa 48 BCE.

During the Ptolemaic period the first Greek translation of the Tanakh—the Hebrew Bible—was undertaken, the Septuagint. It was also the time that some of the esoteric writings attributed to Hermes were being collected, writings and ideas that would have a profound effect on the growth of Gnosticism. Hermes was the Greek version of the Egyptian god Thoth or Tahuti, the god of writing, of mathematics, and of wisdom generally. In his Greek incarnation as Hermes he was responsible for the famous Emerald Tablet upon which was inscribed a short text that is one of the basic documents of alchemy. This text— with its famous dictum "that which is below is like that which is above, and that which is above is like that which is below, for the performance of the wonders of the one thing"—has been translated, pored over, and explicated by a variety of scholars, academics, occultists and alchemists since the discovery of its Arabic version sometime around the seventh or eighth century CE. That it is associated with alchemy is no accident; for generations it was believed that the word "alchemy" was derived from the Arabic *al khemia* or "the Egyptian matter." Indeed, many of the oldest Western texts on alchemy can be traced to Egypt and the Middle East, such as the works of Maria the Jewess.

The relevance of all this for our study of the Phenomenon will become clear.

▼   ▼   ▼

Alchemy is the project of transformation. It involves the mutation of metals in its most popular sense—changing lead into gold, for instance—as well as the production of the elixir of life: a substance that not only would prolong human life but also cure all illness. The Swiss psychologist Carl G. Jung believed that alchemy represented the process of psychological transformation, what he called "individuation," and neglected the purely chemical aspect. In Asia, alchemy was all of these and more. The language of chemistry was interchangeable with that of spirituality, of what we might call psycho-biology. Basically, they are all correct.

While at first glance this has nothing at all to do with the Phenomenon a closer inspection will reveal many elements in common.

That Egypt was associated with this process is based on a very simple observation. The belief of the Egyptians in an afterlife is central to their whole culture, and the way in which they understood life after death is as a process of transformation. This process takes place on worlds other than the Earth, and in dimensions other than ours. The topology of the Egyptian afterlife—the Underworld—is simultaneously here and not-here. The Egyptian god Osiris—the dead and resurrected being *par excellence*—is locatable in a specific time and place, but also in an eternal everywhere in both time and space. His travel takes place from the Earth to the stars, and often through several levels or planes of existence before he reaches his ultimate destination. This transportation is effected by means of a ship, and Osiris is physically prepared for this voyage in a method known to everyone as mummification.

We have already discussed the "stairway to heaven" and the practice of *merkava* mysticism. We know that the ziggurat symbolized just such a stairway in ancient Babylon and Sumer. But while the Jewish and Babylonian systems were detailed and specific in concept, nothing really compares to the Egyptian sources such as the Pyramid Texts and the Coffin Texts and the famous Book of Going Forth by Day (otherwise known as the Book of the Dead). In the Egyptian system the astronomical, biological, and psychological elements are all carefully integrated into a single methodology for achieving immortality which is precisely how alchemy—the "Egyptian science"—was structured. By going back and looking at these Egyptian documents with the perspective of a potential extraterrestrial or non-terrestrial context we suddenly will begin to understand the real meaning behind them. As Dr. Rundle Clark said in the quote that begins this chapter, "When they mythologized they knew what they were doing."

The Sincerest Form of Flattery

Anyone who has watched a child grow up is familiar with how children play. Girls dress up in their mothers' clothes, put on makeup (inexpertly), wear jewelry, and put on airs. Boys imitate their fathers' profession. But even before gender roles become imprinted there is an even more basic form of play in which children simply mimic actions that they see adults perform. They will feed a doll with a spoon and invisible food; they will open a book and turn its pages, pretending to perform an action they don't even understand or have a word for: reading. In other words, children imitate what they see without necessarily knowing what it is they are imitating.

The same is true of adults when confronted with phenomena they do not understand but which they recognize have importance or relevance to their lives, or which are just attractive to them in some way they cannot put into words. In extreme cases, such as the psychopath, an adult may imitate social conventions for the purpose of survival, of fitting in with protective coloration, without having an innate sense of the feeling that motivates those conventions. In less extreme cases we may unconsciously imitate the body language, facial expressions, and verbal manner-isms of people we grow up around or whom we are trying to impress. In these cases we even may know what we are doing, and why. There is a reason for the imitation, a spe-cific motivation. But the imitation does not flow naturally from our being; it is learned behavior. It is not who we really are.

So it is with the Egyptian Book of the Dead.

The texts that were painted on the walls of the tombs and written on papyrus in such elaborate execution and vibrant colors and in such excruciating detail and specific-ity were not the superstitious ramblings of corrupt priests or gullible nobles. Verse after verse of these texts are filled with references to deities, symbols, actions, levels of aware-ness, astronomical data, obstacles, gateways, personalities both good and evil, all tied together in a single insistent narrative describing the passage of a soul from the moment of the death of the body to its final apotheosis. A genuine apotheosis: the becoming of a god, of an Osiris. And this state of immortality, of survival of the identity that is nei-ther life nor death as we know it, takes place elsewhere than on Earth.

One authoritative article on Egyptian myth states:

> Creation myths in any culture are not intended
> as scientific explanations of the way in which the
> universe came into being; rather, they are symbolic
> articulations of the meaning and significance of the
> realm of created being.[5]

While getting closer to our point of view, this explanation still falls a little short. Statements like these presuppose that we know what the Egyptians (or any other culture) were thinking or intending when they articulated their myths, a sentiment that is not supported by any evidence. Since we as moderns feel we know how the universe came into being we look backwards in time at the creation myths of the Egyptians and make assumptions about their motives, their understanding of science, and their expression of these assumed symbolic articulations. We sometimes forget that we also use symbolic language—in the form of mathematics, scientific equations, and the like—to describe the mechanisms behind observed natural phenomena. While string theory, for instance, may be intended as a "scientific explanation of the way the universe came into being" it is far from an accepted explanation and may one day fall into the category of myth once it is replaced by something else.

As we have said many times in this study, what we know as religion is a groping towards a memory of a real event: an event, or events, so out-of-context that language to describe it did not exist. Tom says that religion is the place where science meets spirituality. Peter says that religion is the place where matter meets consciousness.

Same difference.

Science does not have the language, or the skill set, to understand or even acknowledge spirituality as something other than ignorance, but that was not always the case. In ancient Egypt—as in Sumer—there was no meaningful division between science and religion, or even between politics and science and religion. As Rundle Clark offered above, Egyptian mythology had a dual purpose: on the one hand the organization of the world (science) and the coming of the monarchy (politics), and on the other the origin and development of consciousness (spirituality, or religion); but that analysis is the result of the dualism inherent in modern cultural attitudes. For the Egyptians, we think it is safe to say that they did not view their mythology as having two distinct purposes but only a single overriding purpose that combined all the elements Rundle Clark mentioned. In fact, we think it is also safe to say that the Egyptians did not consider their mythology a mythology at all, but a technology. A machine.

And like the child who tries to make a doll drink from an empty bottle, the Egyptians imitated what they saw and what they knew, just like the other Cargo Cults around the world. They were aware that there was a distinct difference between their actions and the event on which those actions were based, but they hoped or believed that through imitation the essence of the original event would be revealed and effected. One day, in other words, the doll would drink.

From that point of view, let us examine the interstellar technology of Egypt.

The Resurrection Machine

Like the tombs of the early Second Dynasty kings, Djoser's tomb was aligned to the north, to the circumpolar stars. . . . He called it . . . "Horus, the foremost star in the sky." We could not wish for a clearer statement of belief underlying the Step Pyramid: that it was *a resurrection machine* designed to propel its royal owner, Horus, to the pre-eminent place among the undying stars.[6]

For thousands of years there has been a great deal of speculation concerning the pyramids: who built them, when, and why were they built? Even questions as to *how* they were built, as the pyramids are the product of a Bronze Age culture but the only tools available to them were made of stone or copper, and copper is a notoriously soft metal that would not have been as useful as iron for carving blocks of stone. However, even that assumption has been called into question since copper—when treated with a small amount of arsenic—can become extraordinarily hard, and it is known that the ancient Egyptians knew of and possessed arsenic in various forms, such as in orpiment (arsenic sulfide): a substance that also was known to (and prized among) the alchemists.

The absence of an established history of ancient Egyptian science has contributed to a type of cottage industry among amateur Egyptologists who speculate about ancient astronauts building the pyramids or some other sort of extraterrestrial explanation. While we are sympathetic to the curiosity and desire of the "ancient astronaut theorists" we feel that their approach unnecessarily over-reaches the data. As professional Egyptologists solve some of the issues that have received the most attention by ancient astronaut theorists, the end result will be a general but unfair devaluation of the ancient astronaut premise, i.e., proving that one claim is false tends to devalue the entire theory. While not all questions concerning the building of the pyramids have been answered conclusively, and while mainstream archaeologists and scientists will resist any explanation that is not anthropocentric, we feel focusing too much speculation on this particular aspect of the pyramids misses the more relevant data that is usually ignored (by both the mainstream and the alternative approaches). This data is contained in the Egyptian texts themselves and constitutes their own direct testimony.

Once we start from our stated position that the Phenomenon is real and first occurred as an event (or events) in ancient history or prehistory, sometimes referred to as "paleocontact," the reason for the pyramids and the religious ceremonies associated with them becomes clear. The richness of Egyptian iconography and symbolism is such that we have a sufficiently large database to work from for understanding what Rundle Clark called the "origin and development of consciousness" that informed the Egyptian impulse towards the deification of their rulers and eventually of their entire race.

It should be remembered from the outset that Egyptian spiritual practices and texts changed and were modified or expanded from dynastic period to dynastic period. We will start with the earliest dynasty of which we have any records and proceed from there, with the caveat that the earliest documents may reflect a clearer memory of the original event but that later texts are just as valuable for they reflect the ongoing attempt by the Egyptian religious specialists to interpret and come to terms with the memory and knowledge they already possessed. Just as science has evolved in its attempt to understand the nature of the material world, so too has religion (or, at least, spirituality) evolved in its attempt to understand the nature of consciousness and its relationship to the material world.

It is also entirely possible—in fact, probable—that there were intermittent events that adjusted Egyptians' experiences of the Phenomenon just as there are in modern times. They would have modified their beliefs in accordance with new contact events. Of course, there may be a multiplicity of different types of contact from different sources and these would have been subject to interpretation and analysis in comparison to the original event(s). We have to remember that the event(s) would have taken place in the stream of Egyptian history—however it affected or altered that historical process—and would have been integrated within their culture to some extent. The pantheism of Egyptian religion may reflect the experience of diversity of that ongoing contact; but there are certain overarching themes that could be attributable to the Phenomenon itself as it is not obvious or self-evident that they would have occurred to the Egyptians in the normal course of their lives. In fact, the sudden appearance of a highly articulated Egyptian civilization out of its background of nomadic and agricultural existence is one familiar argument supporting an external influence of some sort, one that is not locatable in what is known of human history. In fact, the sudden appearance of human civilization all over the world—after more than a hundred thousand years since the first appearance of *Homo sapiens*—is one of the most compelling arguments one can make in favor of an external, civilizing source: one that, for want of a better term or concept, we call "alien."

A case in point:

In the citation that opens this section, Egyptologist Toby Wilkinson describes the earliest known pyramid—the Step Pyramid of Djoser—as a "resurrection machine." This is perhaps the most accurate description so far of the function of the pyramids and rather conveniently reprises our own theme of "sekret machines." When it comes to *sekret*, there are perhaps few ancient machines that meet that designation to the extent that the pyramids do. This is due to the fact that we, as moderns, have difficulty understanding that a tomb can be more than just a burial place. From a scientific point of view, of course, a grave is just a grave no matter how fancy or elaborate. It is merely a container for a corpse. But to the ancient Egyptians a properly designed and constructed tomb was a gateway to the heavens and a promise of immortality. The body thus entombed had to be prepared in a specific way to ensure its viability in the afterlife. This afterlife is not as vague or undefined a destination as the term implies, however.

The Egyptians were not superstitious, credulous people. Their vast construction complexes, their highly articulated religious, political, military and cultural apparatus would suggest a people who had a firm grasp on reality. Yet they did not believe that death was the end of existence. To call the existence of an individual identity after death an "afterlife" is an unfortunate use of language, for life continues through a series of events among which—to the Egyptians—the life we know is but one iteration in a longer series.

A key element of this concept is how the Egyptians understood the basic components of human life and consciousness. Like the other civilizations we've already studied, the Egyptians believed that life originated elsewhere and not on the planet Earth, at least not in the way we understand the "where" in "elsewhere." In fact, their understanding of how the Earth—and humans—were "seeded" is sufficiently similar to the idea of directed panspermia that it demands comparison. This "elsewhere" is what we have been calling the Other World, because it is not only the origin of life but is also the domain of the non-terrestrial forces we sometimes call the gods. Egyptologists call this the Doubleworld,[7] the domain of the *ka* or the "spirit double." It is a dimension that touches ours but only in specific places, such as a statue or an image of a god or of a deceased pharaoh. The so-called "*ka* chapel" is one manifestation of this in Egyptian tombs. It contains a false door or niche before which offerings are made or other rituals performed. This false door leads to the Doubleworld in a way known only to the priests.

According to the most prevalent form of the Egyptian creation story, the god Atum created all life in the universe through an act of masturbation, scattering his seed throughout space. This seed at first gave birth to another set of gods—a male and female couple—who then mated and produced other offspring, etc. This parallels the understanding of creation within Indian religion where we find Shiva—disturbed while engaged in lengthy intercourse with Uma/Parvati—ejaculating into the cosmos and creating the worlds. We will discuss the theories of panspermia and directed panspermia in Book Three, but for now it is worthwhile to note these ideas because of the very clear intention of their respective authors to show that the created world is not the result of the gods having intercourse and becoming pregnant with the world following the example of human or animal reproduction (as would be expected of "primitive" societies) but that there was a seeding of the world by an individual god in a manner that is counter-intuitive and which seems to signal another type of knowledge and its relevant technology. According to the Sumero-Babylonian traditions we discussed earlier, human beings were created from the blood of a slain god and this may reflect a similar point of view: a single deity, usually male, performing creation out of his own fluids or those of another (usually male) deity. Blood in this case may be a euphemism for semen the way thigh is often found as a substitute word for penis.

Why did creation not take place between a god and a goddess, in imitation of how human beings and most other creatures reproduce? Why this additional level of difficulty or complexity? Because the creation of the physical world—and especially the creation of life—was an event related *to* the gods, for which the gods somehow were responsible, but was not *of* the gods. We are something outside of the divine realm, a creation that was external to it, to the Other World or the Doubleworld. An act of creation that involved a god and a goddess would have produced another god or goddess thus keeping the divine force restrained within a tightly knit circle; an act of creation that involved only one of these deities acting alone would produce something else, something outside the divine realm, a penetration into—or creation of—This World.

The act of Atum—or by Ptah, in another version of the same story—creating everything in a solo act of creation, anticipates the biblical account as well. There we have a single deity performing an act of creation over a six-day period.[8] In the most ancient of the Jewish mystical texts—the *Sefer Yetzirah*—a single deity creates the universe through a series of emanations which later become identified with the ten spheres on the Kabbalistic Tree of Life. These spheres or emanations have a certain type of tangibility, perceivable and knowable by those who have been trained to withdraw their senses from the visible world only to train them on the invisible world—the Doubleworld—that is the domain of the gods. (The mystic's ability to keep alert and functioning during long periods of sensory deprivation anticipates the type of training that astronauts undergo in preparation for space travel.)

There appears to be some connection between this divine force emanating from—or secreted by—the gods and the Egyptian concept of the *ka*. *Ka* is a virtually untranslatable term, not because there are too few indications as to its meaning but rather too many. It is found as the root of words as seemingly unrelated as "magic," "sorcerer," "testicles," "vagina," "pregnant," and "copulate," as well as terms referencing thought processes such as "think about," "thought," and "intend." A common thread running through all of these terms seems to be the concept of magical or occult power: the extension of the will into the Doubleworld, or from the Doubleworld into this one. The association between ideas such as thinking on the one hand and fertility on the other implies an association between consciousness and action, with the references to the sex act being allusions to the most basic unity of consciousness and action where thinking is invisible—a kind of Doubleworld function—and action is the visible, This World, evidence of the invisible thought. The presence of concepts such as magic and sorcery on the one hand and those pertaining to sexuality on the other would have been familiar to such Renaissance-era philosophers as Marsilio Ficino and Giordano Bruno who understood that what they thought of as occult power was related to human sexuality and romantic love: all that was missing was the intention of the magician, and the concept "intend" has the same root in Egyptian, *ka*, as the others mentioned above.

Every human being possesses a *ka*, which was usually depicted as a perfect human form, eternally youthful. Its hieroglyphic symbol is evocative: a pair of upraised arms, often shown on the top of the head of a pharaoh, similar to a pair of horns. It is the *ka* that is the life force of a human being, hence this constellation of meanings around it. One definition of the *ka* is "the immaterial, psycho-spiritual aspects of a human being."⁹ Resurrection after death required the union of one's *ka* with one's *ba*.

The *ba* is roughly equivalent to the western concept of the soul. It was depicted in its earliest form as a human body but by the Middle Kingdom it had acquired a very specific iconography, that of a small bird with a human head and face.

The *ba* has an important role in the Coffin Texts, an earlier form of what has been called the Egyptian Book of the Dead. In the Coffin Texts the *ba* is identified with that of Osiris—the dead and risen god—in order to facilitate the resurrection of the pharaoh. There is some disagreement as to the nature of the *ba*: is it part of a composite of body, *ka*, *ba*, *akh* and shadow, the five components believed to be inherent in the human body? Or are these not separable components—some representing the physical nature of the person and others the spiritual nature—but rather inextricable elements of the individual human being? We know from some texts that Horus sends his *ba* to Osiris to enable him to overcome obstacles to immortality. According to some scholars the *ba* is a physical entity, enabling the corpse to walk and even to copulate among human beings. In fact, this idea that there are superhuman beings upon the earth who have gone to the Underworld and come back as living humans, able to perform sexual intercourse with other human beings, seems to recall what we have read about the "sons of god" in Genesis. The Coffin Texts reiterate this formula, constantly referring to the dead returning and copulating with humans.[10] These are, moreover, the dead who have become immortal through the process of mummification and the rituals associated with them, and who have traveled from the Earth to the realm of the stars.

> "I am this great Ba of Osiris, by means of which the gods have commanded him to copulate, which lives by striding by day, which Osiris has made of the efflux which is in his flesh, of the seed which came forth from his phallus, in order that it . . . may come forth on the day on which he copulates. . . . I am the living Ba within his blood. . . . I am the one who, distinguished in his appearances, opens the gates . . ."[11]

The gates are those at different levels of the Underworld—the *Duat*—that are cognate with the different gates or levels in other forms of Middle Eastern mysticism, including both the Sumero-Babylonian forms as well as the Jewish mystical practices known as *merkava* or *hekhalot* mysticism.

Death, to the Egyptians, was the great adventure. It was a voyage as well as a transformative and evolutionary process. It would seem, from reading the Coffin Texts and the Pyramid Texts, that the *ba* of a human being became energized after death especially when it united with its *ka*, or life force. When this happened, the *akh* is made manifest.

The *akh* is the ability of the deceased person to act effectively in This World. It is also a quality of living persons who cause positive change to occur in their society. A living pharaoh could have *akh* just as much as a dead one, but after death the appropriate rituals had to be performed to energize or animate the *akh*.

The shadow—*sheut*—was believed to carry some of the essence of the person since it followed one during life. As in many other cultures, the shadow also carried with it an association with death.

The *ka, ba, akh,* and *sheut* were all present in the human body, the *ha*. Other elements included the heart—*ib* or *ab*—and one's name, or *ren*. All of these were addressed in one form or another in the elaborate funerary rites of the Egyptians. The ability of the dead to walk, to speak, to breathe, to eat, and even to copulate was taken for granted but only if the required rituals were performed. This implies a level of interdependence between the living and the dead that is considerably more intense than those found in many other cultures. The fact that the pharaohs were fixated on this concept to the extent that huge funerary monuments were created for them—the pyramids—with precise instructions and alignments to the constellations implies an even greater concept of the interdependence of humans—both living and dead—and the stars. While the rituals could be performed expertly on the Earth, without attention to the astral structure of outer space the resurrection process would not be successful.

To reinforce this understanding even further, the all-important "Opening of the Mouth" ceremony is evidence that the circumpolar stars held the same importance for the ancient Egyptians as they did for the ancient Chinese and the ancient Indians. In fact, a version of this same "Opening of the Mouth" ceremony was performed in Babylon where it was employed in the animation of the statues of the gods. Recent research suggests that it was used the same way in Egypt—for animating statues and coffin lids (with their engraved or painted faces of the deceased)—and then expanded to include mummies. Some of the implements used in the rituals suggest an imitation of childbirth procedures including a device—similar to scissors—that was used to sever the umbilical cord. Another was a knife in the shape of a serpent (somewhat reminiscent of the *keris*, a wavy-bladed knife common in Indonesia and Malaysia that has mystical associations).

The device that appears most specifically and carefully limned on the panels in the tombs that illustrate the Opening of the Mouth ceremony is one in an odd shape, like a hook. It is an adze and it is in the shape of the Big Dipper (called the "Great Adze"). Referred to as the thigh of Set, it indicates the constellation of the same name that revolves around the polestar.

There is abundant evidence that the Egyptians saw resurrection as involving two different constellations: Orion with its companion star, Sirius, and what we call the Big Dipper. Orion rises and sets and was associated with Osiris by the Egyptians. The Big Dipper does not set but revolves endlessly around the polestar. The former represents the passage of Osiris through the Underworld; the latter represents the approach to the ultimate realm of immortality.

The mummified pharaoh has its mouth "opened" by the priests using the adze, which was made of meteoric iron. In other words, this was iron that had descended to the Earth from the heavens in the form of a meteorite. It is believed that the stone of the Ka'aba in Mecca is made of the same type of meteoric iron.

One of the peculiarities of some meteoric iron is the presence of magnetite: iron with magnetic properties. It is not known if the Egyptians had any awareness of magnetism or that a magnet points north and south. It certainly appears as if the use of magnetic needles in compasses for navigation did not arise until much later, with the Greeks. Therefore, it is curious why the Egyptians would have used such a perfect substance as meteoric iron for the manufacture of this all-important ritual implement fashioned in the shape of that very asterism that points to the North, the place of immortality.

In addition, the construction of the pyramids—including the oldest known pyramid, that of Djoser at Saqqara—was designed with an orientation towards the polestar, as Toby Wilkinson has informed us. The "resurrection machine" involved both mummification—with Djoser as the first known pharaonic mummy, even though his body was never found beyond that of a single mummified foot—and celestial orientation. The orientation had to be celestial and not terrestrial since there was no knowledge of the Earth's magnetic poles at the time the Step Pyramid was constructed.

There is an important issue with regards to our knowledge of the thought processes of the ancient Egyptians. The only written record we have of the earliest dynasties is that found in the tombs themselves. The earliest papyri records are those of a Fifth Dynasty mortuary complex, so again we are dealing with tombs and the organization and running of the tomb complex. Mathematical and medicinal papyri written in heiroglyphics also exist but they appear to date later than the Fifth Dynasty and, anyway, are usually couched in a mythological context as well. Thus we have a tendency to view the ancient Egyptians as a kind of death-and-immortality cult because that is the only information we have. We do not know how they developed their idea that immortality was possible, or that it involved a celestial component that was so important that huge stone monuments—the pyramids—were aligned with it. We can propose that immortality is a basic human desire, but that begs the question: Why?

It is not known whether other mammals or other living creatures harbor a similar desire. We know that the urge to procreate is in all living things to some extent, and that an avoidance of death seems to be a common feature of animals, insects and other creatures as well as of people.[12] While other creatures develop social networks—bees in their hives, ants in their hills, prides of lions, flocks of geese, etc.—they do not create buildings with a view towards resurrection, at least not so far as we are aware. It is too easy to dismiss all this as wish-fulfillment or superstition, which are not really ways of explaining the phenomenon but of marginalizing it. How does one marginalize the idea that gave rise to the pyramids, however, or to the elaborate science of mummification? At some point in their history, we suggest, the Egyptians had evidence that these technologies were viable.

Compare, for instance, the modern fascination with space travel, a fascination that may very well define our civilization the way pyramids and mummification defines that of ancient Egypt. We are told that sending probes out to the farthest reaches of space is important and necessary. We feel that sending human beings to the Moon, Mars, and elsewhere in the solar system is not only feasible but desirable. Yet we cannot articulate the reasons why we would expend so much time, effort, money, calculation, and the best minds of our generation towards this goal. We—and all those countries involved in some form or another in the space race, including Russia, India, China and many European nations as well—take it for granted that we should do this. There may be budget battles in our respective governments over how much money to spend on these missions, but there is no question of shutting down our space programs for good. In fact, private industry has taken up the challenge and there are now satellites in space that were financed by corporations and not governments, and plans for privately owned and privately operated space shuttles and lunar missions.

The phenomenon of global programs for space exploration is the modern version of pyramid building. There is no doubt that we do obtain knowledge from our space programs. We are aware of the possibilities of extracting natural resources, rare metals, and other benefits from planetary exploration. There have been many side benefits from our previous space programs including technological advances that were made due to the requirements of manned space flight. These programs have required that entire societies get behind them, finance them, support them with scientific and engineering solutions, and applaud them when they are successful. Few individuals actually go on these missions; few of us are actually able to experience what it's like to view the Earth from space. We take it for granted that it's wonderful, awe-inspiring, desirable, and even necessary.

We take it all on faith.

Centuries from now our descendants may very well look back on this period the way we look at ancient Egypt. The Egyptians had technology, mathematics, medicine and surgery, cohesive societies, powerful armies and strong governments. They traded with their neighbors. They wrote inspirational texts for their citizens, works of ethics and morality. And they built huge monuments that have lasted for five thousand years. Yet we question their sanity or their intelligence when it comes to their "resurrection machines." And we question the sanity of those who would suggest alternate explanations.

Pyramidiots: A Brief, But Necessary, Digression

In 1968, the book *Chariots of the Gods?* was published to worldwide attention. Written by the Swiss hotelier Erich von Däniken (b. 1935), it claimed that many of the world's most famous ancient monuments—including the Egyptian pyramids—were built by ancient astronauts, i.e., aliens. He claimed that the event in the first chapter of the Book of Ezekiel (which we discussed previously) should be taken literally as a description of a flying saucer, etc. In fact, much of what we have been discussing in this volume could be understood as an attempt to revisit that argument from a more mature and reasoned perspective, one based on a deeper reading of the relevant texts from these ancient cultures rather than an attempt to "decode" (arbitrarily) their architectural forms and technology. We feel that some of the religious experiences as documented by these civilizations are attempts to describe the Phenomenon, but perhaps not in the way von Däniken understood. Nevertheless, his publications on the subject inspired an entire industry of "ancient astronaut theorists" and an enormous speculative literature that often plays fast and loose with the data.

This type of speculation has led to many outrageous claims, such as those by William Dudley Pelley—the leader of a Nazi organization in the United States in the 1930s called the Silver Legion and usually referred to as the Silver Shirts—who wrote that the dimensions of the Great Pyramid could be used as a kind of calculating device for predicting the future. Pelley was a famous believer in UFOs and flying saucers and when he was released from federal prison after the end of World War II he began his own saucer cult (as we will see in the next book). Based on the work of British Israelite David Davidson (who was a member of the Silver Legion), Pelley performed his own pyramid calculations and announced that the Jews would be destroyed on September 17, 2001: a prediction that seems a little startling today when we remember the attacks that took place a week earlier, on September 11, 2001. Indeed, Erich von Däniken's editor on *Chariots of the Gods?* was Wilhelm Utermann, a well-known Nazi author and editor who produced dozens of Nazi-themed publications during World War II under the NSDAP imprimatur and worked as well on the staff of the *Völkischer Beobachter*: the official Nazi Party newspaper. It cannot be denied that the lure of an extraterrestrial hypothesis is linked—in some circles—to ideas of racial superiority and the legends concerning secret Masters lurking in hidden cities. This has led to some declaring a willingness to cooperate with these Masters—whether purely terrestrial "secret chiefs" or largely incorporeal space aliens—in their perceived mission of dominating the planet and "purifying" it of unsatisfactory genetic material. (We will explore this phenomenon in more depth in the third book of this series.)

Mainstream Egyptologists look with scorn on people like von Däniken and Pelley, people they call "pyramidiots": those who speculate about the real purpose behind the pyramids without any actual evidence from archaeological or textual sources, or who write of space aliens or arcane practices that enabled the Egyptians to erect the pyramids and the Sphinx using extraterrestrial technology. Much of this scorn is well-placed, of course. Amateur attempts to answer these questions are almost always extremely speculative, or based on faulty interpretations of key documents or on "channeled" material. There is also a certain degree of intellectual dishonesty on the part of some theorists who ignore evidence that is contrary to their speculations. When professional Egyptologists question these conclusions they are accused of being part of a cover-up, or worse.

However, it is also a matter of record that our knowledge of life in ancient Egypt and particularly the beliefs and understandings that motivated the grand architecture for which they are famous is spotty and often speculative, at best. This has allowed educated guesses by serious independent Egyptologists such as Robert Bauval or Graham Hancock to come up with finely articulated critiques of mainstream opinions on these matters to the extent that they have run afoul of those with a vested interest in maintaining the status quo, among them the famed guardian of the Egyptian "mythos" Zahi Hawass, former Egyptian Minister of State of Antiquities Affairs, and astronomer Ed Krupp of the Griffith Observatory in Los Angeles and a well-known author of popular books on astronomy.

The feud between Zahi Hawass and Robert Bauval is decades-old and is based on Bauval's claim for an earlier construction date for the Great Pyramid at Gizeh and the Sphinx, known as the Orion Correlation Theory. It is Bauval's contention that the three pyramids at Gizeh were built in alignment with the three stars of Orion's belt and that the specific degree of declination of this asterism is a clue as to the era in which the pyramids were erected: a point in time thousands of years earlier than the generally accepted date of the mid-third millennium BCE, or about 2550 BCE, at the beginning of the Old Kingdom. (For comparison purposes, the Step Pyramid of Djoser was built about 2630 BCE in the Early Dynastic period, eighty years earlier.)   Clearly, when Bauval writes of dating the construction of the Great Pyramid to about 10,500 BCE he is challenging the chronology not only of Egypt but of the Middle East in general and with it that of the western world. Indeed, recent discoveries at Gobekli Tepe—among others[13]—suggest that our understanding of the chronology of ancient civilizations does deserve a serious review, but Bauval's Orion Constellation Theory has led to many (sometimes visceral and personal) attacks. Zahi Hawass has accused Bauval of being a Zionist with a hidden agenda, even though Bauval himself was born in Egypt (Alexandria) of Belgian Catholic parents; the accusation involves the charge that Bauval—by claiming the earlier construction date for the pyramid complex at Gizeh—is suggesting that people other than Egyptians built them.[14]

The controversy between Bauval and Hawass is instructional for this project as it illustrates the resistance mainstream professionals employ against those outside their profession. In many cases, as already mentioned, this resistance is justified. One cannot expect those who have spent their entire careers working in a specific field to be open to those from outside who question their judgment. On the other hand, new perspectives that take into account previous research and documentation but which may challenge the accepted wisdom should be entertained without political or personal animus. Not every crackpot theory needs to be seriously and patiently addressed, otherwise that is all the professionals would be doing; but theories such as Bauval's Orion Correlation Theory—based as it is on exhaustive research in the fields of both Egyptology and astronomy—require a more thoughtful critique. However, even the scientific community is not above personal attacks, jealousy, and arrogance and it will circle the wagons against intruders who question their assumptions. This is as good an overview of official reaction to the Phenomenon as any, and evidence for this bias can be found everywhere consensus reality is challenged.

The feud between Hawass and Bauval exploded in April 2015 in Cairo, when Hawass stormed out of a presentation by Graham Hancock in its first few minutes because Hancock had mentioned Bauval's name, causing a minor cause celebre as Hancock tried to reason with Hawass but the Egyptian archaeologist was having none of it. The video of that confrontation went viral among ancient astronaut theorists, as one may imagine.[15]

Zahi Hawass may be a special case. After all, he has been a defender of Egypt's antiquities and a champion of Egyptian national pride in its history and this is to be commended in an era when ancient monuments increasingly are under attack or threat from fundamentalist sources. What this means for Egyptology, though, is that Hawass—who controls access to the archaeological sites—considers himself the arbiter and supreme judge of what is, and is not, legitimate in the field. The downside of nationalism is suspicion of the foreign, the different, and the people and ideas that do not conform to acceptable standards.

However, the astronomer E. C. Krupp also weighed in heavily against the Bauval theory, and for a time it seemed as if the Orion Correlation Theory had been debunked. Krupp applied his skills as an astronomer to the same problem and claimed that Bauval's math was wrong and that the pyramids were not aligned to Orion's belt the way (and the time) Bauval said they were. For some time Krupp's response was considered the last word on the matter.

Then, a number of astronomers around the world critiqued Krupp's critique, calling it "unfounded" and claiming Krupp had "fallen into the trap" of trying to fit ancient beliefs into modern science. The BBC even went so far as to edit a broadcast on this controversy, amending the segment involving Krupp's critique.

However, at the heart of all of this controversy and backlash from the scientific community there was a book published in 1976 to worldwide acclaim that recast the von Däniken phenomenon of the 1960s in a new and much more legitimate light. This was Robert Temple's work, *The Sirius Mystery*.

Robert Temple has the nearly unique position among the alternative science theorists of being a respected and peer-reviewed scholar. A Fellow of the Royal Astronomical Society, and a colleague of the esteemed Sinologist Joseph

Needham, with degrees in Sanskrit and Oriental Studies, Temple's academic and establishment credentials are impeccable. Thus, when his research led him to the hypothesis that an African tribe called the Dogon somehow possessed impossible astronomical knowledge of the binary nature of the star Sirius—sacred to the ancient Egyptians and probable focus of the so-called queen's chamber of the Great Pyramid, as pointed out by Robert Bauval, who actually gained his inspiration from reading *The Sirius Mystery*—knowledge that modern science did not have until the late nineteenth century, and then only with modern telescopes, his book caused a firestorm of controversy. His proposal was attacked on very shaky grounds. Some critics said that the Dogon came by their knowledge from European astronomers who were in the region in 1893 to track a solar eclipse; in other words, the Dogon picked up specific information concerning the binary star nature of Sirius and immediately incorporated it into their religion and culture: a claim that seems rather more farfetched than Temple's suggestion that they came by this information from an advanced (and possibly extraterrestrial) civilization. Others complained that the Dogon seemed to have knowledge about Sirius but did not have knowledge of other stars in the galaxy or detailed information concerning the planets in our system, thus rendering their knowledge irrelevant! It is like the old joke about the dog who could play chess, but wasn't so smart since he lost two games out of three.

Temple got his inspiration for researching *The Sirius Mystery* during a series of conversations he had with Arthur

Young.[16] This is an important name to remember as we will discuss Young at greater length in the next book. As a member of the original "Nine"—human representatives of the Nine Principles and Forces who communicated with Andrija Puharich back in 1952 and 1953 and who would eventually identify themselves as nothing less than the Egyptian Ennead, the Council of Nine Gods that were the supreme rulers of the pantheon—Arthur Young's influence over a generation of serious seekers after esoteric wisdom was profound.

Temple began working as acting secretary for Young's Foundation for the Study of Consciousness in 1966, at a time when he had already known Young for five years. It was during Temple's academic work at the University of Pennsylvania from 1961 to 1967 where he took his degree in Sanskrit. He worked with Young while the latter was writing the books for which he would become famous, but decided in October, 1966 to move to England, and after that time saw Young less often although they remained in contact and visited when they could.

The Sirius Mystery was one result of this friendship. The story of Robert Temple and his work on the Dogon, on ancient languages, the pyramids and, more recently, the Sphinx is beyond the scope of this chapter. There is one point worth mentioning now—in the context of the Phenomenon—and that is that officials at NASA were interested in the book. He received encouragement from the late Captain Robert Freitag (1920-1998) who was assistant director of the Advanced Programs Department at

NASA and who was development manager for the Saturn program and worked on the Mercury, Gemini, Apollo-Soyuz and Spacelab missions from 1963 to 1986 as well. Temple writes fondly of Bob Freitag who had a genuine interest in his work, and who also introduced him to one Baron Jesco von Puttkamer (1933–2012) a German-born engineer who was invited to join the NASA program by none other than Wernher von Braun.

Von Puttkamer came from a prestigious and noble family. A relative—the identically named Jesco von Puttkamer—was a devoted Nazi who had fought for a Freikorps brigade in Munich in 1919–1920 and who later rose in the ranks of the Party and became a liaison officer for Hitler himself. The Jesco von Puttkamer who joined von Braun at NASA was born in 1933, in Leipzig, so he hardly had time to become a Nazi and during most of the war lived in Switzerland with his family. However, he took a dim view of Temple's work and famously wrote a letter on NASA stationery attacking the book and Temple personally. He complained that Temple was talking about "little green men" and that nothing he had to say should be taken seriously by any publisher. This was obviously at odds with Bob Freitag's opinion on the matter, and eventually von Puttkamer recanted—slightly—in a follow-up letter to Temple's German publisher.

Yet, it was this same Jesco von Puttkamer who became a consultant for the first *Star Trek* film and who was a regular at *Star Trek* conventions. His career at NASA was outstanding and he received numerous awards in recognition of his work. But on the question of *The Sirius Mystery*, he and Bob Freitag could not see eye-to-eye and, indeed, von Puttkamer went out of his way to torpedo Temple's reputation in Germany.

In fact, Temple also came afoul of members of the US intelligence community as he writes in *The Sirius Mystery*. Why American intelligence officials should take an interest in a subject as arcane as the Dogon tribe, the star Sirius, and ancient Egyptian religion is a mystery itself and one as yet to be explained. Yet, there was another connection between Temple's work and at least one intelligence officer that Temple himself seems not to have noticed, and that is the work of Peter Tompkins.

Tompkins's first major published work—*Secrets of the Great Pyramid* (1972)—was an immediate success and contributed to the mystique around Egyptian religion. An essential aspect of Tompkins's work was the dimensions of the Great Pyramid, which demonstrate that the Egyptians had advanced and sophisticated knowledge of the size of the Earth, of which the Pyramid was seen to be a representative in miniature. If deliberate—as Tompkins insists—then the technology represented by the Pyramid takes on a grander character and can be seen to emphasize the "celestial voyage" aspect we have been discussing. If the Pyramid is intended to (a) represent the Earth in miniature form due to its proportions and (b) is situated at a precise spot on the Earth's surface meant to emphasize that relationship, and (c) is aligned with specific stars or regions of the Milky Way galaxy, then this "resurrection machine" needs to be examined still further. There is an implication inherent in the structure of the Pyramid that is best explained by the alchemical dictum mentioned previously, that "that which is below is like that which is above . . . for the performance of the wonders of the one thing." The transformative aspect of the Pyramid would be based on its identification of the Earth in relation to the circumpolar stars, and on the sarcophagus as a stellar module of some kind: a point of tangence between This World and the Doubleworld. The wealth of mathematical, geometric and astronomical data in the work by Tompkins contributed to Robert Temple's work on Sirius, and he mentions the Tompkins work several times in glowing terms.

What he does not mention (and may have considered irrelevant) is the fact that Tompkins was an American intelligence agent for the OSS—the Office of Strategic Services—during World War II. He was stationed in Italy, behind enemy lines, in that capacity and was later invited to join the CIA, which he declined to do. His co-author on another book—*The Secret Life of Plants* (1973)—was Christopher Bird, an intelligence agent who worked for CIA and who served in Vietnam and later with the Rand Corporation. The premise of this book is that plants have consciousness and even demonstrate qualities that are sometimes attributed to "quantum consciousness" such as the ability to communicate with other plants at a distance.

This involvement of intelligence agents in the type of arcane subjects we have been discussing is not an isolated case, as the next book will describe in greater detail. Their interest in Ufology is often dismissed on the assumption that the government was concerned over possible political motivations behind some of the more outspoken UFO experiencers, but there is another aspect and that is the possible application of esoteric technology—sekret machines—to real-world situations, as we shall see.

▼   ▼   ▼

We have taken time to look at these particular controversies over the Gizeh pyramids to demonstrate that the antagonism that exists between professional, mainstream academics and scientists and their non-professional, alternative critics often has less to do with science than with personality. Writing about the Phenomenon is a career-ender for many; accusations that bedeviled Temple's work—and that of Bauval, Hancock, and so many others—often were based on the "little green men" association. There is a stigma where this subject is concerned. Bringing it up means you are credulous, not serious, perhaps a little unstable. Writing about it, defending any aspect of it, and you are ostracized, or worse.

Scientists are human, science is not; and trying to fit human beings into a state of pure scientific consciousness is doomed to failure. This is something to be kept in mind as we explore the official reaction to the Phenomenon in the next volume.

Hamlet's Mill

> But nothing is so easy to ignore as something that
> does not yield freely to understanding.
>
> — Giorgio de Santillana[17]

Around the time that *Chariots of the Gods?* was published, another volume dealing with similar material—this time co-authored by a professor at MIT and a German anthropologist and professor of the history of natural sciences—entitled *Hamlet's Mill* came to the bemused attention of the general public. Inspired in part by the *Anacalypsis* of Godfrey Higgins (who proposed the existence of a secret worldwide religion that he called "pandeism" in this work published in 1836) and in part by the theories of Leo Frobenius (1873–1938), the German archaeologist and ethnographer who popularized African art even as he claimed that the lost continent of Atlantis was to be found in Nigeria, this work attempts to show that the world's mythologies are *all* based on the observance of astronomical phenomena and most especially the precession of the equinoxes.

The precession refers to the fact that the spring equinox—for instance—falls on a certain day every year but due to the wobble of the Earth on its axis the degree of the zodiac in which the sun rises on that day is slightly different from year to year. In other words, the sun appears to move backwards through the zodiac, taking about 2,160 years per zodiacal sign. The spring equinox is traditionally the first degree of Aries; due to the Earth's wobble, however, it now actually takes place early in the zodiacal sign of Pisces. It will take place in the last degree of Aquarius (moving backwards through the signs) at roughly 2600 CE, thus ushering in what is called the Age of Aquarius. Traditionally, the discovery of the precession of the equinoxes is attributed to the Greek astronomer Hipparchus (190–120 BCE), and thus thousands of years later than the construction of the earliest pyramid.

The contention in *Hamlet's Mill* and similar studies is that this precession was viewed with some alarm. As the Earth wobbles on its axis, the point in the sky that marks true north changes with the centuries as well. At the traditional time of the construction of the earliest pyramids, therefore, the North Star would have been Thuban in the constellation Draco; today it is Polaris. Thus, according to the theory, the change from Thuban to Polaris was greeted with apprehension and a number of myths were created in response to this event as a means of explaining it or interpreting it in some way, or simply encoding it.

Naturally, this assertion was criticized almost immediately on the basis that there is no physical evidence that could be employed to buttress this claim; rather, what evidence does exist in the form of stone monuments, stone circles, and the like, can be explained without resorting to a sophisticated understanding of the precession of the equinoxes. In fact, according to the detractors, there is no quantitative evidence to show that ancient peoples were even aware of the precession or that it mattered to them in any tangible way.

The precession of the equinoxes takes place over thousands of years and would not have been as dramatic and noticeable as, for instance, a meteor shower or an eclipse of the sun. Further, the bulk of *Hamlet's Mill* seems to involve linguistic and philological analysis that was dated and therefore somewhat suspect, as well as idiosyncratic interpretations of world myths. Thus, even though the book is the work of a distinguished professor at MIT, Dr. Giorgio de Santillana, and his colleague, Dr. Hertha von Dechend, the book's organization, methodology and conclusions were all equally attacked by their academic peers.

However, upon closer inspection, it would seem that the attacks were based not only on pure scientific grounds but also on the affront to the status quo the book represented. Indeed, the case for an ancient tradition involving the precession of the equinoxes cannot be said to have been proved by *Hamlet's Mill*, just as the case for an extraterrestrial hypothesis was not proved by *Chariots of the Gods?* The value of both books lies not in the scientific case they present and the proofs offered, but in the pure concepts themselves. Had the respective authors been a bit more circumspect in their approach to the material they might have found a more willing audience among the circles they most wished to influence.

For instance, a key concept in *Hamlet's Mill* is the identification of myth as scientific information in coded form. In other words, myths can be handed down from storyteller to storyteller while preserving secret knowledge unknown even to the storytellers or to most of the listeners.

This was considered by the authors to be a pre-literate form of scientific language, a method they claim was employed by virtually every culture everywhere in the world.

Objections to this assertion usually take the form that each culture must be studied individually and each culture's myth seen as a product of its distinctive context, and that to draw conclusions based on superficial similarities of theme, symbol sets, etc., is to commit the grave sin of universalism.

While this objection has merit, of course, it can be taken to extremes. This academic fashion has made it increasingly difficult to draw any conclusions about humanity as a whole since it is much safer to contextualize material completely and without reference to other cultures. In its desire to rid the academic world of the type of rampant speculation and the racist, colonialist conceit of the early twentieth-century Austrian and German schools of ethnology and eugenics, modern scholarship—with its emphasis on individual cultures as unique in and of themselves—this academic fad may have inadvertently contributed to a kind of Balkanization of knowledge. Interpretations of data that suggest anything other than independent inventionism are criticized as "diffusionist": a sin remarkably similar to, if not identical with, universalism.

In the claim that the language of myth is a coded scientific language, the authors of *Hamlet's Mill* rightly are criticized for claiming its universality and its ancient pedigree: not because there is a preponderance of evidence against it, but because there is little or no evidence to support their views. However, the language of myth *has* been used to communicate scientific information and since at least the first century CE if not earlier. We find this in the "intentional language" or "twilight language" of Indian alchemy, for instance, as well as in Daoist alchemical texts from China. This method was also employed by Western, European alchemists and using many of the same symbolic references. As Westerners we may ask which came first: the myth or the science? In other words, were the myths developed as a means of communicating secret or scientific information, or was the coded language developed on the basis of pre-existing myths? *Hamlet's Mill* insists that the myths were constructed around scientific, especially astronomical, observations. The problem with this chicken-or-egg approach is that it ignores the third possibility: that both the myth and the language developed concurrently which could be the case if an experience with the Phenomenon is what instigated them. The need for a specialized means of transmitting this arcane experience would produce both the myth and the coded language thus satisfying both context and communication.

A major, if not central, problem where this discussion is concerned is the fact that we do not understand consciousness or the conscious processes involved in acquiring knowledge of the world. We, as moderns, tend to categorize information in discrete and unconnected units for ease of use in scientific study and the application of scientific knowledge. These categories began as largely arbitrary ideas, such as the way we order the classification of living things from kingdom and phylum, class, order, and family to genus and finally species: a system inherited from eighteenth-century observers based on physical appearances, or morphology. With Darwinian evolution and the discovery of the genetic code, taxonomy has undergone substantial modifications in order to accommodate the new data and new ways of interpreting the old data. That's science: it's supposed to change with every new discovery as long as those discoveries fit within the measurable, quantifiable limits of direct observation and the test apparatus developed to aid in their observation. Further, new scientific theories must take into account previous scientific observations; if they challenge those observations then there must be a preponderance of evidence—not intuition or speculation—to support the paradigm shift.

It's a lot like the narration of a story. All the elements have to fit the plot; you can't have the characters switch places or dead people not really dead and living people not really alive without a lot of explanation, enough to satisfy the listener or the reader that there is an underlying logic to it all: that it answers the question put forward by the text. Thus, when amateur archaeologists engage in wild speculation and present it as research or discovery, the rest of the world groans in frustration. It's like a bad plot in a poorly crafted novel.

Our experience of the world, however, is not cut and dried. Science has given us wonderful toys to play with but it has also given us the nightmare scenario of a single person with access to a nuclear weapon or a chemical or biological weapon committing mass murder. We can organize matter, probe subatomic particles, send lasers into space, but we cannot know very much about consciousness even though it is consciousness that is at the root of all technology. Consciousness, according to some philosophers, cannot know itself. It reminds us of the problem science has with the Phenomenon: since it cannot describe it or know it to any reliable degree, it is better to ignore it.

Which is what we do with death.

As technology has made us more and more independent of our society—for instance, able to stream movies into our homes without going to a video store much less a movie theater—it has made us more and more vulnerable to death at a distance. As we hold smartphones in our hands someone else is using the same model to trigger a bomb. Death has become impersonal. We live in independent little cocoons, surrounded by Wi-Fi and drones, oblivious to the possibility that even though we live in an isolated fashion we may very well die in a mass extinction, cheek by jowl with our friends, our neighbors, and total strangers.

But what of our subsequent resurrections?

Have our incredible technological advances—running at breakneck speed through the past hundred years—contributed to an improved method for achieving Egyptian-style immortality?

Can we build a better resurrection machine?

Walk Like an Egyptian

> All of this, I repeat, seems to me curious, obscene,
> terrifying and unfathomably mysterious.
> — James Agee, *Let Us Now Praise Famous Men*

The Step Pyramid at Saqqara is the first Egyptian pyramid. It is the physical boundary marker of the Egyptian dynasties. While Narmer is considered the first pharaoh and is assigned to the First Dynasty, he and his immediate successors were buried in mastabas: large tombs to be sure, but tombs nonetheless. They had none of the grandeur of the pyramids and, more importantly, little of the astronomical and ritual associations so familiar from books, movies, and speculations about Egypt and Egyptians.[18] This all began in the Third Dynasty, the period that ushered in the Old Kingdom and which led eventually to the Great Pyramid of Khufu, in Gizeh, in the Fifth Dynasty.

Why this sudden interest in building pyramids? Why this application of enormous resources towards what was, after all, just a place to bury a corpse? Why did Djoser feel the need to erect this massive structure? Can you imagine doing the same today, no matter how important or powerful the human being? This was a little more elaborate than, say, stuffing Jimmy Hoffa into an oil drum, or that scene in *Goodfellas* where Ray Liotta, Robert DeNiro and Joe Pesci are burying a mob hit in the New Jersey countryside: *"Hey, what do you like, the leg or the wing?"* In our time of cinematic deaths by the millions and novelistic murders and deaths in the millions more, death has lost some of its supernatural aura and burials are largely an afterthought, something to be gone through and not the focus of too much attention: a service relegated to funeral homes and overcrowded urban cemeteries.

In our era of democratic institutions, it is hard to imagine any one deserving such a massive mausoleum, but they do still exist. Hugo Chavez—the controversial president of Venezuela—has been preserved in a sarcophagus like a pharaoh. So has Kim il-Sung of North Korea, and Ho Chi Minh of Vietnam.

And in Moscow and Beijing there are large tombs for Lenin and Mao, respectively, and while they are nowhere near the size and grandeur of the simplest Egyptian pyramid they are impressive by modern standards. In both cases the bodies have been preserved and rest in specially designed and maintained sarcophagi, like pharaohs. The Mao Mausoleum is oriented on a straight north-south axis, but the Lenin Mausoleum entrance faces northeast. The Lenin Mausoleum is notable in this context for it deliberately was designed in imitation of Djoser's Step Pyramid. This odd fact should give us pause.

It can be agreed that ancient Egypt in the time of the pharaohs was a totalitarian state, like the Soviet Union and the People's Republic of China in the twentieth century. Lenin and Mao were the pharaohs of their time and place. But they were communists, which by definition means they were atheists. The mausoleums, through whose doors millions have passed, and their carefully pre-served bodies in crystal coffins have nothing to do with religion as we understand it. One can make an argument for secular religion,[19] but that seems beside the point when we are dealing with such an overt display of bodies that have been dead for decades (in Lenin's case, almost one hundred years). In countries where religion and reli-gious sentiment are seen as counter-revolutionary and decadent, the erection of mausoleums such as these—*and the preservation of the bodies*—suggest there is another purpose at work. Unconsciously, perhaps, those in posi-tions of authority decided that the physical shells of these men should survive forever. In Lenin's case, there was the added precaution of a tomb that was built according to Djoser's design.

In fact, it was the discovery of King Tutankhamen's tomb in 1923 that inspired the Soviet leaders to create a similar, everlasting artifact and was probably the reason why the tomb was designed to look like the Step Pyramid. It was to be a monument to Soviet science as much as a demonstration to their own people of the longevity of the Revolution and its most important leader.[20]

These are impulses that are just as strong in commu-nist and violently anti-religious regimes as they were in ancient Egypt with all its talk of gods and the Underworld and immortality. If we strip the dogmas and the theologies from the mere facts of the preserved bodies and the func-tional mausoleums what is left is a technology. Or, as Lenin himself put it:

All human conceptions are on the scale of our planet. They are based on the pretension that the technical potential, although it will develop, will never exceed the terrestrial limit. If we succeed in establishing interplanetary communications, all our philosophical, moral and social views will have to be revised. In this case, the technical potential, become limitless, would impose the end of the role of violence as a means and method of progress.

— V. I. Lenin, in a conversation
with H.G. Wells, 1920[21]

For those who may be concerned that I am overstating the significance of the Lenin angle, I submit the fact that the first ever radio broadcast directed at a potential extra-terrestrial civilization was the November 19–24, 1962, transmission to the planet Venus by the Soviet Union. The message was in Morse code and consisted of three words: MIR (or "peace" in Russian) transmitted on November 19, LENIN, and SSSR (which is the phonetic English equivalent of the Russian CCCR or in our spelling USSR), both transmitted on November 24, 1962. (One wonders how the Venusian were supposed to know Morse code, but it was a start.)

For years after his death, Lenin was eulogized heavily. He was often identified with the stars in the sky, or as a guiding star looking over Russia, or as a "new sun" in the heavens,[22] concepts that would be very familiar to the ancient Egyptians. Of course, much of this is typical poetic imagery but remember the degree of censorship in the Soviet Union and the fact that all print media was subject to government control and supervision. At the same time, the Russian Orthodox Church was being suppressed, the cathedrals closed or turned into museums or government buildings, and priests imprisoned or forced into hard labor, all while the "secular religion" aspect of Lenin-worship was approved by the state, as was—of course—his preserved corpse and the step pyramid mausoleum.

To make the transition from mere human to an immortal god complete, in 1932 the Russian artist Kazimir Malevich offered his sculpture of Lenin atop a skyscraper designed as a spaceship, pointing to the heavens. "Here was the ultimate parabola of the revolution—Lenin as rocket."[23]

Imagine for a moment the reaction of an ancient Egyptian priest to the Lenin Mausoleum and the preserved corpse of the communist leader. He would have recognized the construction of the tomb, of course, since it is a deliberate imitation of Djoser's. He may have been confused as to its orientation, however, and wondered to which star in the heavens it was aligned. He would have been astonished at the degree of preservation of the corpse which far exceeds anything the mummification process could produce, until he looked more closely and realized that the body was not mummified but embalmed and that the integrity of the body—its viability as a device for entering immortality on the long voyage to the stars—was damaged.

Lenin's body had had its arteries severed and the blood drained from it, and that meant that pumping formalin through it in the normal embalming procedure was not possible. Formalin—the combination of formaldehyde and alcohol or water—was injected instead into various organs of the body, a less than optimal solution. Every two years or so the body is removed from its sarcophagus and the whole process repeated, with the result that less and less of Lenin's body is preserved; a plastic compound is used to compensate for those areas that have not survived. This, to an Egyptian priest, would have been unacceptable. The mummification process preserved as much of the original body as possible, and the embalmers of Lenin's body were determined not to allow mummification to take place. They were seeking an alternative that would allow the body to be viewed as if it was sleeping rather than dead: a cosmetic agenda rather than a resurrectional one.

We have taken the time to describe the differences and similarities between the Lenin "Step Pyramid" and the original in order to emphasize the purpose of the Egyptian "resurrection machine." The appearance of the body was immaterial to the Egyptians; what was important was that the body was intact, at least insofar as its musculature and skeleton were concerned. Naturally, they could not leave the heavy organs in place so they were removed and placed in what are called canopic jars. These jars, four in number, were hermetically sealed and placed next to the mummy at the time of its internment. The body itself was dried out with natron and then covered in bitumen and other substances that would retard decay for a very long time. The idea was to remove as much water as possible from the body. This had the (unexpected?) benefit of allowing the DNA to survive much longer than would be possible otherwise, enabling genetic researchers today to recover DNA from the body of, for instance, King Tutankhamen. Further, high levels of radioactivity have been discovered in the pyramids, leading Robert Temple[24] to suggest that this also may have contributed to the amazing state of preservation of the mummies. Embalming fluid, on the other hand, accelerates DNA decay and it is only recently that some methods have been devised that would make it possible to recover DNA from embalmed tissue.

Thus we have to ask why the Egyptians went to all this trouble when it would have been easier to simply bury the bodies the usual way. They may not have intuited the necessity of preserving DNA so that the body could—at some later date—actually be cloned, but they did understand that it was necessary to preserve as much of the original body's capacity for walking and talking (and copulating) as possible. The organs were similarly preserved, so was it with a view towards enabling some future scientist-priest to "reassemble" the body and reanimate it?

As mentioned above, reanimating the body was the reason behind the Opening of the Mouth ritual. The Egyptian priests were not stupid; they knew that the corpse in front of them was not going to get up and walk suddenly or speak with them. The ritual had an enabling function, however, and that was to encourage or permit the conscious element of the individual to take up residence in the body. If the body's mouth was "opened" in this world then it was enabled to speak and eat in the next. The body of the deceased, properly prepared, was a tangible link between This World and the Egyptian Doubleworld. The priests performing the funerary rituals (and all the rituals that took place later in front of the *ka*-chapel, rituals that continued to "feed" the *ka* of the pharaoh and otherwise maintain it) were Mission Control. The pyramid was the launch vehicle. The pharaoh was the astronaut (literally, a "traveler to the stars").

And the pharaoh—as a mummy—was in hibernation for the long journey home.

Sleep Like an Egyptian

> A stairway to heaven shall be laid down for him,
> that he may ascend to heaven thereon.
> — *Pyramid Texts*, Utterance 267

Space travel has its own peculiar requirements and effects. As the "twin problem" of relativity demonstrates, any sort of travel close to the speed of light would result in a strange circumstance. The twin who traveled aboard a spacecraft for, perhaps, two years before returning to Earth would find that he had only aged those two years but his twin—left behind on Earth—would have aged thirty years. This is demonstrated by the Special Theory of Relativity which deals (among other things) with the way we experience time.

What if this had occurred in ancient Egypt? Let us say that there had been contact between the Egyptians and an extraterrestrial with a device for space travel that included a ship and a launch platform. The ET decides to take an Egyptian on a trip to the stars, to inspire (or terrify?) her. In order to do so, the ET places the human in a state of torpor, i.e., a kind of suspended animation or hibernation for the long trip. The other Egyptians see the process, or parts of it. They try to fit what they are seeing into some context with which they are familiar. Their friend appears to be dead, but they are assured that she will come back to them. She is contained within a glass coffin. There are strange markings on the coffin.

Politely, the ET asks the Egyptians to leave as he makes preparations for take-off. They stand back and watch as the ET enters the device. They can't see what is happening behind the walls and doors of the launch platform but there is a shimmering of some kind and then everything goes still.

And that's it for thirty years. She is believed to be dead after all this time. Her parents grow old and die. Her friends age. In thirty years a lot has happened.

And then.

The launch platform begins to shimmer once again as it had so long ago. A door opens, and she steps out looking exactly the same as when she left. But her parents are gone. Many of her friends have died or moved to some other city. Those who know her are amazed. She is no longer human. She can't be. She has not aged in thirty years. She has been frozen in time. Her twin sister, bent with age, gray-haired, bewildered, reaches out and touches her sister's face, gently, in awe. There is only one explanation.

She has become immortal.

The old priests check their notes from thirty years ago, unfold the papyrus scroll, check their calculations. A "dead" Egyptian woman. A mysterious coffin. A vehicle for traveling to the stars. A specific direction with regard to known constellations in the sky. And immortality.

They look at each other, and at the lady. And back at each other. A raised eyebrow. A shrug. A gleam. A half-smile.

"It's worth a shot," they agree.

▼   ▼   ▼

While fanciful and a little dramatic, the above scenario is one way of accommodating what we know of the Egyptian funerary rites and pyramid architecture into a logical narrative. Our technology is starting to imitate the speculations of the "pyramidiots," from putting our astronauts into a chemically induced state of torpor where they rest in glass sarcophagi for months or years—like mummies in their tombs—to the general acknowledgment that the pyramids (and even the mastabas before them) were aligned with celestial phenomena.

These are just the bare bones of the process, however (no pun intended). The complexity of the instructions, prayers, and other symbolism represented by the Pyramid Texts and the Book of Going Forth by Day indicate that the Egyptians understood the terrible dangers in this type of "travel" and of the need for the traveler to be in perfect spiritual, as well as physical, condition. The psychological makeup of a modern astronaut is a subject of great concern to NASA and other space agencies, as important as their intellectual and physical condition. To remain in a spacecraft for extended periods of time—for instance, aboard the International Space Station for months or longer, or a Mars mission that may require years of living in space—requires a certain degree of psychological toughness, courage, and reserves of inner strength.

One would imagine that none of this would be required of a pharaoh's corpse, but due to the Egyptian concept of the *ka*, the *ba* and the various other components of a human being, the death of the body did not imply the complete eradication of identity. Death became a leaving of the Earth in the sense of Earth as a stand-in for This World; while the physical remains were left "here" like discarded clothing or a butterfly's cocoon, the rest of what constitutes personhood survived in the liminal space between This World and the Doubleworld. It is this very liminality that is dangerous, both to the living and the dead. The continued communication or interaction with the dead in the form of ritual observances, offerings, and the like, reinforced the memory of the dead among the living (the way it is usually explained) but also provided a kind of substance or corporeality to the dead, empowering them to function in this liminal space. It was believed that the dead underwent a series of trials on their way to immortal status, the way Catholics used to believe in Purgatory: a liminal space between this world and heaven (the sky) where sins were purged until the soul was cleansed and purified. Among the ancient Egyptians, this Purgatory state was a series of levels on the way to the circumpolar stars. By participating in this voyage—by assisting the dead pharaoh to make his way through the many levels of the Duat, the Underworld, by reciting the right spells and making the right offerings, *on the right days* (implying a certain degree of coordination with the dimension of time and its associated celestial components)—the Egyptians rehearsed their own after-death experience.

This is paralleled in the Tibetan scripture known as the *Bardo Thodol*: what popularly has been known as the *Tibetan Book of the Dead*. The Tibetan system anticipates a future incarnation, however, and not a state of immortality among the stars, but we can regard it as a kind of functional immortality since the essential element of the individual returns, again and again, to inhabit a series of bodies. The most famous reincarnated Tibetan is, of course, the Dalai Lama whose successor is chosen among children born at a specific time (astrology being an element of this process) and who can identify the previous incarnation's possessions and recognize his friends. The deceased lama is guided through the various levels of the Underworld—called the Bardo, in this case—by the other monks who perform rituals and recite the appropriate scripture, enabling the spirit of the lama to pass from one stage to the next. The Bardo is the intermediate stage between death and rebirth, and the recitation of the Bardo Thodol is supposed to help the lama in this intermediate stage as he listens to the instructions of the monks.

In New Age circles, both ancient Egypt and Tibet are considered spiritually advanced cultures and attract tens of thousands of seekers and interested laypersons every year to museums, conferences, seminars, ritual initiations (in the Kalachakra Initiation of Tibetan Buddhism), and sales of bestselling books. These are also the two cultures with a very highly articulated concept of the nature of the afterlife, with its multiple levels and characteristics, and the need of the deceased to have prayers recited over them that match these levels and are designed to guide the soul through each obstacle.[25]

The six levels of the Bardo have been characterized as stages or types of consciousness.[26] If we apply this same approach to the Egyptian texts we may find important new data that will enable us to understand the Phenomenon in terms of its manipulation of consciousness. We may also identify the link between these death rituals from very different cultures and the nature of the paleocontact that we propose inspired the Egyptian funerary system.

The Egyptian understanding of the nature of the Underworld and the identification of the human soul with a divine counterpart underwent changes and modifications from dynastic period to dynastic period. In the Third Dynasty, the soul of the deceased king was identified with Horus, the falcon-headed god who was the son of Osiris and Isis. It is only with the Fifth Dynasty that we begin to see a shift from Horus to Osiris with the deceased king identified as "an Osiris," i.e., a dead-and-resurrected god. That may not be particularly relevant to our investigation as the basic form still applies: the identification of a human being with a god, a process that takes place after death, and the elevation of that human being to the celestial realm.

However, since Osiris was identified with the constellation we know as Orion, the orientation of the Great Pyramid to this constellation reflects this changing view. To be sure, only one of the shafts of the Great Pyramid is oriented to Orion; the other shaft is oriented to the circumpolar stars of the Northern horizon, which maintains the original focus of the Third Dynasty Step Pyramid: the constellation we know as Ursa Major but which to the Egyptians represented the god Set who was predominate in the early dynasties. Thus what we may see in the Great Pyramid is not so much a modification of the original concept but an elaboration of it, a refinement that recognizes the polarity of these two celestial objects and their contribution to the resurrection machine.

More recently it has been suggested[27] that the shafts of the Great Pyramid were not intended to align with either Orion or the polestar specifically but were designed to form a passageway from one part of the Milky Way to another.

The Milky Way was known to the Egyptians as the Winding Waterway, often associated with the goddess Nut. Egyptologists are divided as to whether Nut represents the Milky Way or just the path of the ecliptic. She is usually portrayed stretched out over the Earth with her head in the

West (where she devours the Sun every evening) and her womb in the East (where she gives birth to the Sun every morning). She is usually depicted nude and covered with stars.

There are many references to the Milky Way—as Winding Waterway—in the Pyramid Texts. There, the dead pharaoh is said to cross the Winding Waterway to take his place among the circumpolar stars. This would seem to lend credence to the idea that the shafts in the Great Pyramid were aligned to capture the light of the Milky Way to provide the "sea" on which the sarcophagus would sail. Either way, the scholars agree that the shafts are there to provide a passage to the stars. Thus, the pyramids have a celestial component and it is not possible to understand ancient Egyptian religion (inseparable from ancient Egyptian science, mathematics, and architecture) without it.

The so-called Tomb of Osiris illustrates this concept perfectly. A replica located 114 feet deep below the surface of the earth in the Shaft of Osiris at Gizeh, it is a plain stone sarcophagus on an island surrounded by the waters of a man-made canal. There are many strange characteristics of the Shaft of Osiris, which are noted by Robert Temple,[28] including some quite ancient sarcophagi located on the second level of the three-level construction. The Tomb was subject to some devastation and vandalism in the post–World War II era with its four columns or pillars destroyed and the sarcophagus damaged as well, but the overall design of the Tomb is clearly evident even today.

This is a feature that is largely unknown in Western architecture aside from the simple orientation of Catholic churches which usually have the altar in the East. Asian architectural forms often represent beliefs surrounding celestial associations, such as *feng shui* in China and *vaastu* in India. Of course, it has been shown that ancient Western monuments such as Stonehenge in England as well as some of the earthen monuments left behind by the Mound Builder culture in North America are also celestially oriented or aligned. What usually goes unnoticed in these cases is the possibility that they, too, were designed with the idea of celestial resurrection in mind. Normally, edifices like Stonehenge were believed to be erected solely for the purpose of marking the seasons through solar and lunar observations and possibly as sites for ritual observances, but there has been little thought given to the possibility that they also were used to transport the dead to the heavens.

Egyptologists often characterize the pyramids as a refinement of the Primeval Mound concept in Egyptian religion. According to one of the Creation stories, a mound appeared in the midst of the Primeval Ocean and from this mound the god Atum emerged to begin the process of Creation. In other recensions, Atum is himself identified as the Mound. Thus the Mound already had spiritual and narrative significance for the ancient Egyptians that perhaps predated the erection of the first pyramid at Saqqara.

William F. Romain makes just this point when he writes of the "azimuths to the Otherworld" he discovered among the orientation of burials in Mound Builder sites in North

America.[29] His findings reveal that many of the Hopewell sites are aligned to points of solar and lunar significance, and that the burials were often oriented to the Moon and the western sky. Unfortunately, the Mound Builder culture did not leave behind any written records and all that remains are the mounds themselves and whatever objects and artifacts have been discovered buried within them. The orientation of many of the sites can be determined, however, and they are startlingly consistent in their attention to astronomical alignments, whether these were burial sites as already noted or mounds of ceremonial significance (such as the famous Serpent Mound in southern Ohio). It is sobering to realize that most Americans are totally unaware of the remains of these ancient cultures right beneath their feet, remains that are in some cases (the Adena culture) older than even Solomon's Temple in Jerusalem.

The stratification of the Egyptian Underworld—the Duat—probably did not take place in any formal way until the Eighteenth Dynasty when the Book of Gates became known. This text shows the Underworld as having twelve gates which the deceased must pass before attaining immortality among the stars. The Book of Gates appeared long after the Great Pyramid was built, long after the end of the pyramid era. It reflects in a textualized form what the pyramids did in architecture and is basically a commentary on the earlier funerary practices that gave rise to the pyramids themselves.

The number twelve is probably more of a symbolic device, representing the twelve hours of darkness or night, than it is a literal view of the nature of the Duat. However, it could also be seen as the imposition of ideas of order and symmetry onto the more primeval concept of the Underworld and this type of refinement may reveal a deeper (or, at least, more contextual) understanding of what the Underworld is composed and how travel through the topology of the afterlife is achieved. It is possible— let us entertain the idea, anyway—that early impressions of the Underworld were that it was a chaotic place whose symmetry or order was not immediately apparent, reflective of a human inability to determine the lines and planes of that Otherworldly dimension, but that with continued exposure to it, the vague outlines of a system were gradually understood and perceived.

The deceased pharaoh first would face the prospect of securing passage to the Underworld in a craft of some kind. This is depicted as a boat that would carry him and a number of gods and assistants on the journey through the twelve gates.

At each gate he would be confronted by two gods and a serpent. He would have to know the names of these three in order to pass through the gate. (This system has persisted down the centuries; it is reflected in the similar method known to *merkava* mysticism of passing through the seven levels of the heavens to reach the throne of God. Each level has its name and its guardians, and these also must be known in advance.) These levels contain all sorts of obstacles, such as lakes of fire, fire-breathing serpents, and, of course, the dreaded Apophis: a giant serpent that was the adversary of Ra. Even Set—considered by this time to be the enemy of Osiris and Horus—would be enlisted to fight Apophis during the pharaoh's journey through the Underworld. It was believed to be the blood of Apophis that colored the sky red in the morning and in the evening as Set wounded him while protecting Ra's solar barque. (This may be a survival of the cult of Set that had persisted in Egypt until the worship of Osiris became the dominant religion.)

At the fifth gate, the deceased's heart is weighed. If it is heavier than a feather of Maat—the goddess of Truth—then the deceased is devoured.

And so it goes. Why did the Egyptians anticipate so many hardships and dangers *after* death? Why did they emphasize the need for meticulous preparation of the body and the soul of the deceased? In short, why were these rituals needed and what were they imitating?

We have already pointed out the visual similarity between a corpse in its sarcophagus and an astronaut in the state known as torpor, a kind of hibernation during long space voyages. The funerary rituals also refer to the *ka* of the pharaoh and his *ba*, uniting after death to create the *akh*: a combination of both the energy of one and the image and personality of the other.

This may anticipate various technologies, everything from robots to avatars and the internet of things. The idea of an extension of consciousness through various media—essentially traveling to space in virtual transports—means that an individual may operate equipment at a distance:

on the other side of the world, or on the Moon or another planet. At the time of this writing there is concern among the American military at the advances being made by other countries in the development of robot armies. In addition, drone technology has made it possible for a young operator sitting in a control room in the United States to manipulate a drone flying over the mountains of Afghanistan in real time, seeing with the drone's cameras and firing missiles from the drone's wings. Would the ancient Egyptians have interpreted a drone as a *ba*: a half-human, half-birdlike creature flying between the tomb (the pyramid) and the Underworld?

The authors of this book have communicated with each other using a variety of media, none of which existed during the Third Dynasty. We can send e-mail, use text messages or cellphone calls, and even use video conferencing such as Skype. Is the image I see on Skype something real? Peter knows it is not Tom in front of him, physically, but he can see him and hear him speak. In other words, it is Tom's double that is present on the computer screen, his *ka*: Tom disassembled into electronic impulses and then reassembled on the other side of the country. We do this, thinking we understand the science and the technology behind it, but of course we do not. We are using tools that were developed by others—largely nameless and unknown to us—whose basic theories of operation involve everything from Newtonian electromagnetism to quantum physics, manipulating waves of light and sound over vast distances.

Tom has appeared in many music videos. These videos have streamed and been downloaded all over the world. A century from now, these moving images of Tom will still be visible and audible. To an ancient Egyptian, this would mean his *ka* still exists, his double. In other words, that Tom has achieved immortality. Not as a metaphor, but—to an ancient Egyptian—in reality.

Take these different technologies, all of which have come into their own in the last thirty years or so, some of them more recently than that, and some still evolving. Space flight. Artificially induced hibernation. Electronic transmission of sound and images as well as control functions across continents and across space. The constant attention to astronauts in space from ground control. And the "twin problem" of the Special Theory of Relativity. All of these were imitated and anticipated in the pyramids of Egypt.

Are we going to space, or are we building resurrection machines?

And what if it's both?

The Opening of the Mouth ceremony. Notice the priest holding the adze that is in the shape of the seven stars of the Big Dipper, or "Thigh of Set." Immortality was linked to this asterism in many cultures around the world, for the Big Dipper rotates endlessly around the Pole Star, the "Throne of Immortality."

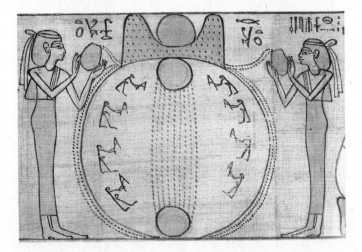

The rising of the Sun at the moment of Creation. Notice the two female figures each pouring water into the ocean that encircles the Earth, while Sun rises in the center between two mountains or hills. Compare to the following:

This is a Tarot card from the *Tarot de Marseille*, and shows a woman pouring two vases of water into a river. Above her are seven stars, plus an eighth in the center. The center star, larger than the others, has eight rays which, in Sumer, represented the "heavens" and "divinity."

An image of the *ba*, the Egyptian soul, in gold. The ba is a winged creature that is sometimes depicted as rising from the corpse of the Pharaoh during the mummification process.

Zahi Hawass, the man who has been in charge of Egyptian antiquities for many years, and who has clashed with those who claim an older date for the building of the pyramids and the Sphinx at Gizeh.

Jesco von Puttkamer, a NASA scientist and German-born engineer who was brought to the United States by Wernher von Braun. He met Robert Temple, the author of *The Sirius Mystery*, through an introduction by a mutual friend, Bob Freitag of NASA. Strangely, von Puttkamer did what he could to stop *The Sirius Mystery* – which is about the knowledge the Dogon tribe of Mali, Africa, have about the physical properties of Sirius, the Dog Star – from being published. Temple's book is concerned with Egyptology as well as archaeoastronomy, and while Freitag was enthusiastic about the work von Puttkamer was horrified by it. He is not to be confused with a relative of the same name who was a devoted Nazi during the war.

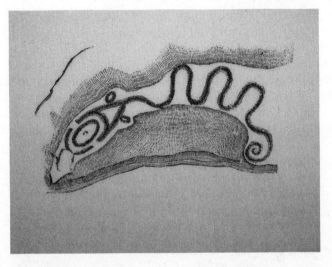

The Serpent Mound in Chillicothe, Ohio: an ancient earthworks that depicts a serpent swallowing a pearl or a sphere. It is over two thousand years old, and the largest serpent effigy in the world.

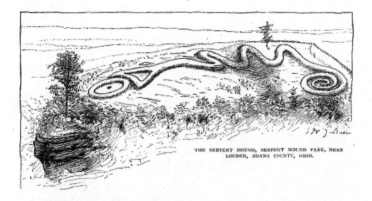

THE SERPENT MOUND, SERPENT MOUND PARK, NEAR
LOUDEN, ADAMS COUNTY, OHIO.

Another view of the Serpent Mound. Adena and Hopewell mound constructions often had astronomical as well as spiritual functions.

Compare this to:

This was the flag of the Qing Dynasty in China until 1912. The Dragon and Pearl motif is common throughout China and the Chinese diaspora worldwide and can be found on textiles, porcelain, carved wood, paintings and sculpture.

This is a pyramid, covered with grass and trees, that is the tomb of Emperor Jing of the Han Dynasty, located outside of Xi'an in China.

This is Lenin's Mausoleum in Moscow. It was designed specifically in the style of the Stepped Pyramid of Djoser in Egypt, and Lenin's body preserved in a specially-constructed sarcophagus like a modern-day Pharaoh.

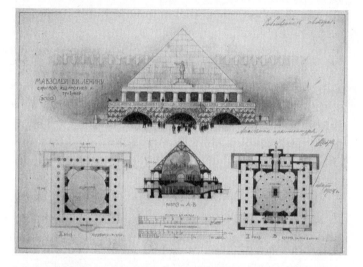

These are architectural studies for Lenin's Tomb which show details not easily seen from outside the edifice. The intention was to duplicate as closely as possible the design of the great Egyptian pyramids.

In Beijing, the body of Communist leader Mao Zedong also is preserved in his own mausoleum.

THE STAR WALKERS

The North is the area of the Great Yin that presides over death. It is located at the deepest bottom of yang and also found deep within the earth . . . . The underworld is situated at the lowest point of the earth, which however reflects the highest point of heaven and is the exact mirror image of the realms-on-high. There is an essential unity between the underworld and paradise . . .[1]

I N THE PREVIOUS CHAPTER WE READ OF THE STRANGE symmetry of the ancient Egyptians who saw travel through the Underworld and eventual arrival at the region of the circumpolar stars as two aspects of the same after death journey. They can be likened to the two shafts of the Great Pyramid at Gizeh, one pointing to the North and the circumpolar stars—especially the Big Dipper—and the other to the South and the constellation Orion with its companion star Sirius. The "setting and rising" aspect of the journey reflects the daily motion of the sun and also the setting and rising—the death and resurrection—of the deceased pharaoh, known in the Third Dynasty as Horus but in later dynasties as Osiris. The circumpolar stars, however, do not rise and set but are in eternal motion around the polestar; this is the symbol of immortality itself, a place the Chinese called the "Garden of the Immortals."

In Chinese Daoism (Taoism) this duality of underworld and "realms-on-high" is more formally depicted as mirror images of each other, as complementarities. This may help us to understand the Egyptian concept and, indeed, the idea of celestial flight is intimately connected to the idea of immortality in the Chinese as well as the Egyptian cultures. In Chinese Daoism celestial flight was a complex affair that was accomplished by ritual specialists during their lifetimes with the goal of rising to the stars and obtaining the gift of immortality there. In Egypt it would seem that only the soul of the dead enjoyed the voyage to the stars. Once the Pyramid Texts were painted on the walls of the pyramid the entire edifice was sealed with the king's body in its sarcophagus within it, thus permitting only the soul of the king to read and follow its meticulous instructions. In China, two thousand years later (and perhaps much earlier, if it is indeed a survival of more ancient shamanistic practices), this process was extended to include the living in the mystical practice known as the Pace of Yü, to which we soon will return.

Regardless of cultural differences and the all-important contexts that define individual and social responses to the material it is clear that immortality and celestial flight are connected. For some reason, human beings from many different countries and eras hold the belief that travel to the stars is an essential prerequisite for longevity, as well as healing the sick and the attainment of supernatural powers. This "ecstatic flight" is a hallmark of some of the individuals known as shamans. Historians of religion such as Mircea Eliade and Ioan Couliano made special studies of shamanic ecstatic flight and noted its presence in cultures as diverse in time and place as Siberian shamans, Jewish mystics, and the ancient Greeks.

The Chinese belief that a living person could travel to the stars seems to contradict the Egyptian belief that only the properly prepared dead could do so; however, in both cases, it is believed that what does the traveling is something other than the physical body, even though the physical body may benefit from it in the long run. Both systems describe an inner aspect of human consciousness that is projected outward from the Earth into the empyrean, and which returns with celestial knowledge and superhuman capabilities.

And in both systems gates must be passed and obstacles encountered. Both systems acknowledge that this is a dangerous process and only those psychologically and physically prepared are able to undergo its rigors.

There is another similarity, although it may seem rather opaque at first. It concerns alchemy.

The Chinese alchemical literature contains elements and basic concepts that are strikingly similar to the ones we find in the Middle East and Europe at roughly the same time. The themes of material transformation as well as spiritual transformation are present in both systems, and both are linked to the dream of immortality. The alchemical process is a form of material regression: that is to say, it is a going-back to the original matter, the *prima materia* from which everything evolved, and then moving forward again only at a highly accelerated pace. Both systems are also psycho-biological: the whole body-mind continuum is involved in the process. There are constant references to seminal fluid, to blood, to germination and reproduction, and to breath. In other words, to the basic elements of the Creation stories we have already studied.

If we compare the alchemical systems to those Creation legends we will find that there is a great degree of congruence between such seemingly unrelated concepts as the importance of blood to the Aztecs, the sons of God in Genesis, the creation of human beings in the Sumero-Babylonian texts, and the auto-erotic act of Amun in the Egyptian texts; between the celestial flight of the pharaoh and the celestial flight of the Daoist adept; and the pursuit of immortality as a glyph for transport to the stars.

There can be seen a groping in the dark towards an idea but dimly sensed, that humanity is a creation of beings not of this Earth and that the secret of our salvation is in our apotheosis, an apotheosis which is at the heart of every revolution: the revolutionary who becomes the king. The message in all of these methods—all of these machineries of joy—is that we cannot remain as we are and survive.

Fortunately for the sitting "king," our knowledge is divided against itself. Science is at loggerheads with consciousness, and in that division we are vulnerable to any force or race of beings which have managed to unite these disparate strands into a single club with which to beat us. Our lack of imagination in this regard may prove fatal. We may say that every time science and reason suppress religion, and every time religious fervor overwhelms science and reason, a demon gets his horns.

In this chapter we will discuss the role of consciousness in manifesting, experiencing, and interpreting the Phenomenon and we will use the Asian approach as our lens. The Egyptian approach did not focus as much on consciousness the way we understand it: it was a system designed for the dead. While the Egyptians seemed to believe that the dead possessed consciousness—otherwise the funerary practices would have had no meaning or relevance—the Asian approach involves the fully alive human being. While many of the same technologies are present in Asian culture as in the others we have discussed—such as sacred architecture, alignment with the stars, and celestial flight—the Asian methods focus more specifically on consciousness, although not to the exclusion of the other elements. As we will see in more detail in Book Two, even the twentieth-century father of modern Chinese aeronautics, Qian Xuesen, recognized the value of these ancient systems and their applicability to space flight. As we look back along the timeline to the earliest recorded "myths," and then forward to today we will see that humanity has been traveling along a single arc. At various intervals the Phenomenon has intervened—or has seemed to intervene—either to encourage us in our quest or, as some think more likely, to monitor and perhaps thwart our progress along the way. Perhaps it is only by being who we are—this awkward combination of animal and god—that we humans can survive and perhaps even triumph.

▼   ▼   ▼

In the last chapter we discussed the pyramids of Egypt. Admittedly we did not focus too much on the astronomical alignments or the geodetic aspects of the constructions aside from some generalities. That material is covered in so many other books, some more or less valuable than others, that we did not want to waste space here with a lengthy digression. It is worthwhile, though, to point out that the Middle East is not the only locale that boasts pyramids.

We know that pyramid-shaped structures can be found in Mesoamerica, everywhere from Mexico to Guatemala and even in Peru. What may not be known as well is the existence of large pyramid complexes in China.

While familiar to a handful of travelers in the early twentieth century, they did not come to world attention until World War II when one—the so-called White Pyramid—was spotted by an American aviator with engine trouble. After several passes over the site, which is near the ancient capital city of Xi'an, James Gaussman reported back what he had seen and the news spread. He talked of a shimmering white pyramid-shaped structure with a flat top and what appeared to be a crystal or some other shiny, jewel-like object at its summit. Later, a pilot of TWA confirmed the story and published a photograph of the site that had been taken by Gaussman in 1945.

The structure was later identified as the Maoling Mausoleum, a tomb dating to the first century BCE. It is the final resting place of Wu Di (140-87 BCE), the fifth king of the Western Han Dynasty. Its base is larger than that of the Great Pyramid at Gizeh, but it is shorter. It does not rise to a point but is truncated, giving it the appearance of a trapezoid. No one knows why Gaussman thought the top had a crystal as there is no such object to be seen now, but that does not mean it wasn't there in the past. That area of China was off-limits after the Revolution and was remote at the best of times, with very few foreigners ever reaching that deep into the countryside. In addition, the Chinese government did not want to acknowledge the existence of the pyramids for several reasons.

In the first place, since they did not have detailed information about the structures themselves they could not afford to appear ignorant so it was easier simply to deny they existed, at least temporarily. In the second place, there was no evidence that the structures were built by Han Chinese. They may just as easily have been erected by people other than the Chinese, and that would present a political problem. Indeed, such a situation did take place when it was rumored that one of the pyramids was actually of Turkish origin. In addition, the Chinese ordered that at least one of the pyramids—the so-called White Pyramid—be covered in trees so that it would blend into the landscape like an ordinary hill.

Be that as it may, China does boast numerous pyramids and even has a version of its own Valley of the Kings in the region around Xi'an. There are between seventy and one hundred tombs in that area, many of them in pyramid shapes: either rising to a point, like the Great Pyramid, or truncated at the top. As in Egypt, the Chinese pyramids were tombs for kings, and as in Egypt they were astronomically aligned.

While some are believed to be as old as the Fifth Dynasty Egyptian pyramids, most probably date from the first millennium BCE to the first millennium CE. They have been subject to all sorts of speculation, of course, and some ancient astronaut theorists suggest that they were built by the same mysterious civilization that was responsible for the erection of pyramids all over the world, from the Americas to Asia and everywhere in between. While that theory may seem farfetched we feel it is necessary to acknowledge that there are scholars who claim that there *was* communication between ancient Egypt and ancient China and that the one culture could have influenced the other; or, conversely, that they both stem from the same ur-culture. This is usually predicated on the basis of similarity in architecture, religious or other symbolism, and even linguistics.

What we are looking at here is a topic of some controversy in scientific circles: the debate between diffusionism and independent inventionism. Basically, traditional science supports the independent inventionist position that states there was little or no contact between ancient cultures in geographically distant regions and that any similarity between one culture and another is the result of independent invention: in other words, if both the Aztecs and the Egyptians built pyramids it was the result of each culture coming to that architectural design independently of the other. There was no cross-fertilization or "cross contamination" between Mexico and Egypt.

The diffusionists, however, propose that there was travel—particularly sea travel—between the Americas and Africa, the Americas and Europe, and/or the Americas and Asia before Columbus "discovered" America. They point to a number of anomalies in the archaeological record to support this proposal and, more recently, research in genetics and other sciences has been cited as evidence for at least some degree of diffusion. Scientists and skeptics generally ridicule diffusionism, however, citing a lack of concrete proof of contact between the continents. Ironically, it is perhaps the nature of modern science—the increased specialization that separates archaeologists from geneticists, astronomers from biologists, physicists from linguists, etc.—that tends to support the idea of independent inventionism as each specialist lives and works in relative isolation from other specialists, thus contributing to an "independent inventionist" mindset. There have been, however, exceptions to the rule.

The German Tibetologist Seigbert Hummel (1908–2001) was the author of numerous books and hundreds of peer-reviewed articles on the subject of Eurasia and the similarities between Chinese and Tibetan iconography on the one hand, and Egyptian motifs on the other. Only one of his books was ever translated into English—*On Zhang-zhung*, a treatise on the language of a mysterious kingdom to the west of Tibet that was believed to have been the origin of the Bön religion—but some of his most important articles on the subject of Egyptian-Asian diffusion were published in Italian translation under the title *Tracce d'Egitto in Eurasia* (Egyptian Traces in Eurasia) by Guido Vogliotti. In this collection, Hummel's ideas concerning the dissemination of Egyptian religious ideas and iconography are detailed along with comparative illustrations, not only of images but also temple floorplans, etc.

Seigbert Hummel spent the war years in Leipzig and Dresden as a Lutheran minister while studying Asian languages and Egyptology in Munich, Rostock and Tubingen. From 1947 to 1955 he was the curator of the Asian collection at the Ethnographic Museum of Leipzig, but then left that position to become pastor at a Lutheran church in the German-Czech border town of Vogtland. From then on to the 1990s, however, he continued to contribute to the literature on Sinology, Tibetology and Egyptology, publishing hundreds of articles, mostly on Tibet.

One of his projects was the comparison of the Naxi script—in use by a Chinese minority in the Himalayan foothills—to Egyptian.[2] Some of the Naxi ideograms are remarkably similar to Egyptian both in design and meaning, although not in pronunciation (but then the beauty of ideograms is that they have no intrinsic phonetic component, which is why a Mandarin speaker will pronounce a Chinese character differently from a Cantonese speaker, but will mean the same thing).

Another article in the collection is "Il mito di Osiri in Tibet" (The Osiris Myth in Tibet), an attempt to show that—even more than ideograms or iconography—Egyptian narrative was imported into the Himalayas as well. While it is always interesting to notice similarities in architecture or iconography between cultures it is more convincing when ideas are discovered that seem to reflect a commonality of understanding. That the Egyptian pyramids and the Chinese pyramids were both used as tombs for kings underlines an essential similarity that bears further investigation. If these two cultures also have other ideas in common we can speculate about a common source. Tracing the possible path of diffusion is one way to determine how—and when—one culture influenced the other. If it can be shown that two cultures developed similar ideas at about the same time, however, then diffusion may be less likely and we have to look farther afield.

One such attempt at demonstrating the diffusion of cultural content across continents is the work of Stephen Oppenheimer, whose scholarly work—*Eden in the East*—is buttressed with his own work in the distribution of genetic factors related to various forms of tropical diseases. Oppenheimer discovered[3] that when there was a difference in genetic susceptibility to disease there was a related difference in some essential religious or historical "myths." His work in immunology in Southeast Asia—in particular in Indonesia and the South Pacific—led him to the discovery that even clans that shared the same island could have different genetic profiles and that this difference was paralleled by different creation or flood myths. It was this striking discovery that led him to undertake the type of research into ancient cultures that we already expect to find in von Däniken, Robert Bauval, and Hertha von Dechend, with the exception that Oppenheimer is a medical doctor and an expert in his discipline who has spent considerable time in the field and came by his theories through exposure to various cultures *in situ*. By tracing the path of genetic profiles from the South Pacific to Europe, he was able to demonstrate that the original civilization—the "ur-civilization" that is sometimes identified, or confused, with stories of Atlantis or Lemuria—was located in what the geologists call Sundaland: that area, now largely submerged, that stretches from continental Southeast Asia all the way through the Indonesian archipelago and beyond. The myths of the Great Flood—which many modern scholars tend to see as unrelated stories of different floods taking place at different times all over the world—are shown by Oppenheimer to exhibit a degree of consistency with each other and to represent the memory of one massive inundation that wiped out the Sundaland civilization even as survivors made their way by boat to Africa, India and the Middle East, bringing with them the rudiments of their sunken culture.

Toby Wilkinson, an Egyptologist who undertook several expeditions to the Egyptian desert east of the Nile to examine ancient petroglyphs, came to the conclusion that the Egyptian religious concept of a boat carrying the gods to the Underworld had its origin with the people who made the petroglyphs. In a book[4] published in 2003, he proposed that there was a race of people that evolved into what we know today of the ancient Egyptian civilization, and that their religious concepts formed the basis of Egyptian religion (rather than the "ancient astronaut" theories he rejects, which imply that some sort of extraterrestrial civilization gave rise to the pyramids as well as to the cosmological and mythological concepts that surround them). As a representative example he analyzes the many "boat" petroglyphs that have been discovered in the Eastern Desert as religious artifacts referencing the Underworld. These rock carvings depict ships with oars that are remarkably similar in design to those found on Egyptian papyri and other artwork of the dynastic periods, leading Wilkinson to believe that it was the religious ideas of the Eastern Desert people that contributed to the funerary customs and beliefs of their descendants and that these ideas came from nowhere else. In other words, he assumes that the rock art had a religious function that was distinct from any sort of secular purpose such as a historical record.

However, if Oppenheimer's thesis in *Eden in the East* is correct, these petroglyphs could just as easily refer to actual boats carrying people from an actual sunken civilization (a civilization now in the Underworld, quite literally). There is no need to interpret the Egyptian petroglyphs as religious iconography the way Wilkinson does; to be more precise, there is no reason to interpret the petroglyphs as *solely* religious. Our intention has been from the beginning to dissolve the archaic and dualistic boundaries that separate religious experience and expression from historical accounts. What if the petroglyphs represent a spiritual undertaking that has its roots in an actual event, which, of course, is what we have been saying all along?

Another scientist—with a PhD in geology and geophysics from Yale—has offered further evidence that the pyramid-building culture had its roots in Southeast Asia and spread from there to other parts of the world including Africa, the Americas, and elsewhere in Asia.[5] In fact, in a book published the same year as Wilkinson's, Robert M. Schoch has even conceded that the basic ideas of theorists such as Robert Bauval, Graham Hancock and Adrian Gilbert concerning the astronomical orientation of the pyramids have merit even if there is room to contest their measurements at certain points.[6]

More than that, however, Schoch goes on to demonstrate that there is sufficient evidence linking Mesoamerican culture with Indian and possibly Southeast Asian culture to warrant a new look at human history. Casual observers have often remarked on the uncanny similarity that exists between architectural forms found in the Mayan ruins, for instance, and those found in India and throughout Indian-influenced cultures of Southeast Asia, such as those found in Indonesia and the former Indochina. Schoch contends that the people of Sundaland migrated outward from Southeast Asia sometime between 6000 and 4000 BCE, heading northwest to create the pyramid cultures of Sumer, Mespotamia and Egypt and east, possibly as far as the coast of Peru.[7] The circumstance that precipitated this migration was a Great Flood; however, he also proposes that a cosmic event was involved and possibly responsible for the Flood itself.

We may remember from the chapter on Mexico that the Aztecs reported the source of their culture as a land surrounded by water from which they arrived in Mexico by boat. Many academics have attempted to situate this land to the north of Mexico somewhere in the southwestern United States, perhaps in the midst of a lake. However, if we can entertain Schoch's idea—and Oppenheimer's thesis—that there was a genuine Great Flood that decimated the Sundaland civilization, then the Aztec legends become more than myth and closer to a historical account.

> The myths explain the world as it was and as it came to be; thus, in a very real and concrete sense, myths tell history.[8]

What, then, are the implications for all of this where consciousness is concerned? And, finally, how does all of this impact our understanding of the Phenomenon?

Come Fly with Me

> For in Taoism, as in all other religious conventions, what the meditator experiences and describes as an inner ascent is a reversal of the processes of what

the outer world sees as cosmic 'creation.' The adept personally returns to the source of things.[9]

Ecstatic flight is probably one of the earliest forms of mysticism and even of religion. What we know as shamanism—an early form of spiritual contact—is often associated with ideas of flight to the stars. This type of flight is effected through ritual and through sensory manipulation: the use of drums, horns, hallucinogenic plants or fungi, and other technologies for withdrawing the senses from the measurable, observable world and into an interior world of visions. Impressions and information collected during the experience of ecstatic flight were then applied in the real world for healing, predictions, and other uses.

An essential aspect of many shamanic practices is the experience of death and rebirth; thus, just as in ancient Egypt, death and rebirth are linked to flight to the stars. This is further reinforced by the concept of an *axis mundi*: a pole that runs through the Earth and directly to the sky above. Shamans were known to climb trees or specially designed poles in order to make contact with celestial beings or to enter into the heavenly realms. In China, the shaman was known as *wu*—indicating a person who was able to contact the celestial powers—and held a special place in Chinese Neolithic and later society. In fact, the case has been made that the earliest Chinese leaders were also shamans: that political power stemmed from celestial power.[10]

The astronomical alignments of ancient Chinese burial and temple complexes dating to the Neolithic era are further evidence of this relationship between political authority and celestial contact. This association between the stars, shamans and the state continued for centuries through the Shang and Zhou dynasties, that is, from circa 1560 BCE until about 200 BCE. The *wu* had many characteristics in common with other types of shaman, such as the ability to predict the future, healing of the sick, and knowledge of astronomy and astrology. The methods varied, but usually included entering a trance state through the use of drums and specific forms of dance.

That the Neolithic-era Chinese held many of the same beliefs as later generations to the present day can be discovered in the burial sites and temple areas that reflect the square-and-circle topology of the cosmos familiar to anyone who has studied Chinese religion. In this cosmological system, the square represents the Earth and the circle represents the Sky. The harmonious union between Earth and Sky is the hallmark of Chinese state religion as well as of mystical practices. The Earth is represented in the divination system known as *I Jing* by the trigram for Earth which is composed of three "yin" lines, or broken lines. The Sky is represented by the trigram for Heaven, composed of three "yang" or solid lines. These forms are found at various Hongshan sites, including Niuheliang in the far northeastern boundary of China, in Inner Mongolia.

In China, early shamanic practices became more sophisticated in Daoist (Taoist) forms of meditation and alchemy but they did not lose their essential character. The Daoist emphasis on returning to origins, conquering death, and ascending to the stars is indicative of the themes of the "ur-religion" we have been referencing. Why is a return to origins associated with celestial flight and immortality in traditions as diverse as European alchemy, Indian Tantra and Chinese Daoism? There does not seem to be any logical reason—based on observations of nature, for instance—why this should be so. Yet, we find in diverse cultures around the world this insistent theme that celestial flight is connected to immortality.

If a return to origins can be interpreted as a return to *human* origins, then we are back in Genesis territory and the Sumero-Babylonian Creation Epic. Here we treat the origins of human beings as a result of the manipulation of matter by other forces. This would seem to be the logical association of celestial flight and return to origins, for these other forces—according to this theory—would be the alien gods whose presence gave rise to notions concerning religion, supernatural powers, creation, etc. And the association of these ideas with immortality reveals the impetus towards spirituality that is at the basis of human yearning. Rather than some irrational defect in human consciousness, religion becomes a logical reaction to actual events in dim pre-history. We can take either a purely literal approach—saying that, with Zecharia Sitchin and the other "ancient astronaut theorists," there were actual aliens visiting the Earth who deliberately created human beings as some kind of slave race—or we can take a more transcendent approach and say that our beliefs and yearnings are evidence of our otherworldly origins via panspermia or some other type of "seeding" of this planet, the Earth, with life forms from else-where. We cannot, however, derive conclusions based on an analysis of data that ignores the very real and persistent nature of the Phenomenon in sightings and experiences; thus our thesis leans more towards Sitchin than it does a purely accidental, if not transcendental, explanation without embracing Sitchin's theories completely or abandoning the ideas of directed panspermia, for instance.

Either way, religion—and myth—hold clues to the mystery of human life and act as containers of potential solutions to some of humanity's oldest questions about purpose, meaning, and survival.

Further, if we take the myths concerning origins, flight, and immortality as evidence of real events or real circumstances, then we must also address the persistent stories of a revolt in heaven and a conflict that involves beings other than humans in a struggle that takes place off-world but which has direct consequences for our own. The conflict between Tiamat and Marduk has echoes in other cultures, other traditions that may stem from an initial cause, and this "ur-conflict" re-emerges in one form or another in some of the most iconic struggles in human history as we re-enact that original trauma like serial offenders compelled to revisit their own abuse as children in the crimes they perpetrate as adults.

The history of Daoism usually begins with the story of Lao Ze (Lao Tse) who lived in the fourth century BCE and is credited with having written the *Dao De Jing* (Tao Teh Ching), the classic text of Daoism. However, the Book of

Changes or *I Jing* (I Ching) is believed to have been created around 1100 BCE, perhaps as an outgrowth of Shang dynasty oracle bone divination. While not a Daoist text per se, it is usually associated with Daoism and with the basic themes of Daoist philosophy.

Many people are aware of concepts like yin and yang: the female and male principles underlying existence. They may also be aware of the hexagrams of the *I Jing* and the divination system associated with them. There is, however, much more to Chinese philosophy and especially to Daoism. This is not the place to go into a detailed description of Daoist rituals and beliefs; for that, the reader is directed to the sources given in the bibliography. For now, we will focus on several elements that have relevance to our project.

The cultures of Neolithic-era China boast several features in common with its Egyptian and Mesopotamian counterparts. The Hongshan culture of northern China and Inner Mongolia demonstrates that astronomically aligned monuments—including pyramids—were known in Asia as early as 3500 BCE, if not before. While a large number of tombs, including a pyramid, were discovered at the Niuheliang site in the 1980s, a large temple area—located underground—was also uncovered. In that space, known as the Goddess Temple to archaeologists, several statues were found that were of enormous size. In one case, a statue three times the height of an average human being was discovered, as well as life-size sculptures of human heads. One such head is frequently mentioned in the literature as being of a "goddess," although there is no ancient writing found at the temple that would support that theory. This particular head has eyes made of green jade. That jade was being worked into jewelry that had sacred significance as early as the Neolithic age is already well-known. The images that were carved into jade include dragons and the peculiar cloud-formation designs that would become famous in later Chinese art, specifically related to Daoist conceptions of otherworldly powers and influences. This demonstrates a steady continuum of artistic and religious design from Neolithic times up to at least the Ming Dynasty, and beyond, making Chinese culture as consistent as (if not much more than) Egyptian. Further, there was no common language uniting all of China from the northeastern border where the Niuheliang culture was found to Xi'an—where the famous terracotta warrior tomb is located—and to the vast "Valley of the Kings" outside Xi'an where more than a hundred tombs and pyramids are located, thus indicating the strength of cultural and spiritual ideas that transcended ethnic and linguistic diversity.

The pyramid and tombs at Niuheliang, for instance, are aligned along the north-south axis, just as the pyramids and mastabas in Egypt. This northward orientation would become further refined in Daoist religion millennia later. One of the more popular Daoist deities is Doumu, the "Mother of the Chariot," which is a reference to the Big Dipper. She is the consort of Shangdi, the "Primordial Emperor," and they are both often worshipped together in Daoist temples. The Big Dipper—the circumpolar stars of the northern sky—is called by various names in China, from "Northern Bushel" to "Northern Dipper" and "Great Chariot." Each of the seven stars is given a different name and associations, a practice that is followed in India and in many other Asian countries which see the seven stars of the Dipper as seven gods.

While the pyramids of China and of Egypt were aligned to the polestar and associated with the seven stars of the Dipper, both cultures had extensive rituals concerning death, resurrection, and immortality that accompanied this astronomical orientation. In other words, this was an active process of transformation and celestial ascent and not merely a genuflection in the direction of a possible or promised site of rebirth. Among the Egyptians, this process involved a priestly hierarchy caring for the dead pharaoh and assisting in his ascent to the immortal sky. Among the Chinese, this process became more articulated—possibly due to the resilience of ancient Chinese culture and its continued survival over five thousand years—so that an individual, while still alive, could proceed along the same path as the pharaoh who had to die before the crossing could be initiated.

This process was known as the Pace of the Stars, and utilizes the shamanic dance known as the Pace of Yü.

> Tours of space were a commonplace of ancient China.
>
> — Edward H. Schafer[11]

The ancient Chinese shamans contributed this arcane practice to later generations of Daoist magicians. It consisted of treading on the stars of the Dipper in a specific way. This was done either through a shamanistic trance or, as in later practice, on an actual diagram of the seven stars of the Dipper, embroidered on a carpet or in some other fashion. The "Pace" was named after the Great Yü, the "Divine King" of Chinese legend whose mother had conceived him "after seeing a shooting star in the constellation Orion."[12] (Once again, as we saw in ancient Egypt, we have an association between the Big Dipper and Orion.) The Divine King had a strange, halting step according to the legend. This was because he sacrificed part of his physical body in order to obtain supernatural powers.

Treading on the Dipper was an essential element of one of the ancient Creation epics of the Chinese. The secret method of Treading on the Dipper enabled the Yellow Emperor to defeat Ch'ih-yu, who was a sea monster with horns, eight fingers and eight toes and an array of metal weapons. The name Ch'ih-yu is derived from the word for reptile,[13] *ch'ung*, which further connects this myth to those of Egypt and Mesopotamia. He was both a rain god and the god of war, and eventually gave his name to a form of martial arts, *ch'ih-yu-hsi*.

This Chinese myth has obvious resemblances to Indian and Babylonian myths. We see the world in disorder and the earth covered with water; a fearful monster presides over the water and disorder; he is vanquished by a god who proceeds to put the earth to order.[14]

Thus, the sea monster was defeated using a technology associated with the Big Dipper, a technology taught by the Mystic Woman of the Nine Heavens.[15] The Nine Heavens refers to the Dipper as well; the Chinese astrologers believe that the seven stars of the Dipper are accompanied by two invisible stars, Fu and Bi. The source of this belief is unknown. It may have some connection with the Egyptian idea of the Ennead: the Nine Gods who rule the Egyptian pantheon and who will return in twentieth-century America as literal space aliens aboard a starship, as we will see in Book Two.

> The North is the place of origin, the area of cyclical change where the embryonic stage and the process of gestation are located . . . . The direction is symbolized by the agent water, which is where all life begins.[16]

Where we have only scant information from the Egyptian texts on the importance of the Dipper, especially in the Opening of the Mouth ceremony, the Chinese texts provide us with a richer field of associations. As can be seen from the citation above, the Dipper was connected to water and the origin of life, which emphasizes its relationship to the Creation epics of so many world cultures. Treading the Dipper is the equivalent of the Egyptian voyage through the Underworld, with the exception that it takes place deliberately by living persons. The destination is the same: the polestar, the place of immortality. There are stages in the journey, and deities associated with each stage. In this it is also similar to the Jewish mystical systems of *merkava* or *hekhalot*. In the Chinese case, we have evidence that it grew out of shamanistic practices involving ecstatic flight. It is too early to say whether the Jewish and Egyptian systems also derive from some form of shamanism; there is too much controversy over shamanism itself—what practices may be considered shamanistic, what ethnic groups are the proper locus for shamanism *per se*—but we can suggest that ecstatic flight is the common denominator among all of these technologies.

That the physical structures of the Egyptian and Chinese pyramids and their orientation along the north-south axis give further support to a common understanding of the relationship between a properly arranged burial and immortality seems obvious. Initiatory practices picked up where the more elaborate funeral practices left off. The time of the Han dynasty in China and the Ptolemaic dynasty in Egypt saw the growth of individual mystical practices that built upon the death-and-resurrection practices of the state. Greco-Roman Gnosticism, Mithraism and alchemy were developed in the first century BCE and the first century CE in the Mediterranean countries. In China, organized Daoism began roughly during the second century CE, even though Lao Ze himself flourished centuries earlier. By the time the Shangqing school of Daoism was formed in the seventh century CE—with its sophisticated forms of ecstatic flight and alchemical practices—Tantra had begun to develop in earnest in India and incorporated many of the same ideas concerning immortality, alchemical transformations, the creation of the world, and a form of internalized ecstatic flight. The Big Dipper was also an important asterism in India and the seven cakras or subtle centers of the human body were identified with the seven stars of the Dipper.

All of this is mentioned to emphasize once again the commonality of ideas that transcend geographical and ethnic differences. These are crucial concepts that surround humanity's most pressing existential concern: survival after death and its corollary, celestial origins. And they are interpreted the same way across cultures. In China, we have an additional set of clues as to how this technology functions and they illustrate the relationship of consciousness to the process even more clearly than the Egyptians due to the availability of textual material.

In Daoism, the Underworld and the Celestial Palace are mirror images of each other; in fact, they are the same place seen from different perspectives. That is why death is such a potent metaphor for celestial ascent. The Daoist magician denies his physical body through fasting and meditation, withdrawing his consciousness from the body and even—in some cases—visualizing the organs of his body as external to himself.[17] This is reminiscent of the shamanic experience of being disemboweled and having the organs washed and reinserted into the body to create a new physical structure more in line with the requirements of celestial ascent. We say in the West when someone has died that they have "gone to heaven," but we do not mean this literally. In fact, we have often forgotten why this cliché remains so persistent. In the ancient Egyptian religion as well as in the practices of Chinese shamanism, this concept is meant deliberately if not literally. In other words, there is a direct relationship between death and celestial ascent even though the juxtaposition of those two events is really counterintuitive. They don't make sense when taken together, but they have become so fused in common usage that we don't really think about it. The Chinese systems afford us greater understanding of this relationship by underlining the idea of immortality as something to be gained by first denying the body's natural tendencies towards eating, drinking, sleeping and copulating—in essence, restricting the body's movements and functions as much as possible in order to enter into a state resembling death—and then by moving consciousness towards the celestial Pole in a series of meditations that are tightly structured around each of the stars as gates leading to further development.

The Pacing of the Void

The Dipper . . . is the carriage of the emperor. . . [18]
—Si Ma Qian, Chinese historian, d. 86 BCE.

There are various approaches to Pacing the Void, but they all involve a focus on the stars of the Big Dipper. One approach involves the adept sleeping on a diagram of the Dipper, and then experiencing the Dipper as a chariot carrying the Supreme God—Tai Yi—down to Earth. Another, more complex, involves a circumambulation of the nine stars of the Dipper (the seven visible and the two invisible) in a counterclockwise motion followed by a "treading" upon the individual stars of the Dipper using the Pace of Yü so that at no time do both feet touch the star at the same time. Then there is a reverse treading of the stars, leading to a clockwise circumambulation of the asterism, then another treading of the stars but this time with both feet landing on each star. This process begins by visualizing the Dipper and all of the divinities that reside within each star.

Another version has the adept using a piece of silk on which the Dipper has been embroidered, a silk that must be kept locked away and out of sight of the profane until the ritual begins. At that point the adept is clothed with the stars of the Dipper and rises to the heavens to enter the constellation, saluting the individual deities on each star but approaching the Dipper through its "dark" side first.

These and many other versions are all concerned with treading or pacing on each individual star of the Dipper, each of which has its specific qualities, attributes, deities, etc., and in every case the overriding image is that of a human being ascending into space from the Earth and passing through the seven (or nine) gates that guard the entrance to the polestar: the source of immortality.[19]

> Pacing the celestial network . . . is the essence of
> the flight into the heavens, the spirit of the pace

of the earth and the truth of all movements of humanity.[20]

The above statement from one of the "liturgical texts" of Shangqing Daoism spells out the meaning of the practice in no uncertain terms, although it is usually interpreted in a broadly metaphorical sense rather than as a literal statement of fact. *All movements of humanity . . . the flight into the heavens . . .* could anything be more specific than that? The Daoist adept was not merely a solo operator seeking to "pace the void" for his or her own gratification or enlightenment, but was the advance guard of the rest of the human race in its astral aspiration.

▼   ▼   ▼

Just as we find in the Egyptian Pyramid Texts and the Book of Going Forth by Day, as well as in the techniques of Jewish mysticism, each of the stages along this celestial journey is marked by a specific deity, password, and other associations. It is the passage of the soul through various conditions, each more rarified than the last. Shangqing Daoism, however, is even more specific when it comes to the nature of each of these stages and goes so far as to reveal the existence of an "anti-matter" characteristic of this space voyage.

To Daoist adepts, the seven stars of the Dipper are mirrored in seven "dark stars" that are invisible to the naked eye but which can be seen by the adept during their journey.

These dark stars are silent, but their presence is essential to the overall celestial network along which the adept passes on the way to the polestar. These dark stars are not the same as the two invisible companions of the Dipper, Fu and Bi, which round out the total number of stars in the Dipper to the aforementioned nine. While Fu and Bi may be invisible, they are nonetheless tangible: they are merely unseen. The seven (or nine) dark stars, however, are both invisible and intangible; they can be sensed and "seen" by the adept but only after progress has been made during the celestial ascent.

It is almost too easy to draw comparisons between the dark stars of Chinese astronomy and the dark matter of the physicists. In the Chinese example, however, the dark stars are specific locations, mirror images of actual stars, and actually anticipates modern theories concerning anti-matter. Yet, in late 2007 a theory was advanced that the earliest stars in the universe actually were dark stars.[21] Professor Katherine Freese of the University of Michigan proposed that these first stars were fueled by dark matter, which is the predominant matter in the universe comprising some eighty-five percent of the total.

In some versions of the Chinese cosmological system *all* planets and stars have dark counterparts, which may be a way of proposing an "anti-universe": a (so far only) theoretical construct in which every particle has its opposite, a universe that exists at the edge of our own. This theory states that at the Big Bang equal amounts of matter and anti-matter would have been created, even though virtually no anti-matter so far has been detected. However, contact between a particle of matter and its opposite particle of anti-matter still would generate energy as the mass of each particle would be the same even if their electrical properties (for instance) differed, resulting in mutual annihilation.

Dark matter, on the other hand, comprises about twenty-three percent of the universe. It is matter (not anti-matter) that emits no light, which is why it cannot be seen. Most likely the Chinese concept of dark stars could be taken to represent dark matter. Was this concept developed by the Chinese purely through imagination or speculation? It has been presented as a fact discernible only to those who have advanced along a path of initiation sufficiently to be able to detect it independently. Chinese astrologers make allowances for the dark stars in their horoscopes, and as we have seen, the Daoist sages insist on the presence of two dark stars in the asterism we call the Big Dipper.

The Dogon in Africa are said to be aware that Sirius—the Dog Star, so important to Egyptian religion—is a binary star even though its counterpart cannot be seen with the naked eye. There has been a great deal of controversy as to whether the Dogon arrived at this knowledge independently, via the transmission of information from some other, ancient, source, or from contact with European astronomers; however, the Chinese belief in dark stars is demonstrably ancient and could not have been acquired through contact with western astronomers. Did they simply imagine the existence of dark stars, of invisible matter, of a network of celestial mechanisms from which humanity descended?

At some point the question must be asked: Does a work of the imagination qualify as true? Can one imagine the real?

Naturally, scientists have employed imagination—and even dreams—as auxiliary tools in their quest for knowledge. Niels Bohr and Wolfgang Pauli come immediately to mind, as we will see in Book Three. Leaps of imagination have led to discoveries in every discipline from physics to chemistry to biology. Yet the role of imagination in the development of scientific knowledge is regarded as a kind of accident, and consciousness itself is something that defies scientific study.

The Chinese, however, made the disciplined use of consciousness an essential element of their mystical approach to reality. Imagination, dreams, visions, sudden insights, could as easily be the product of a systematic practice involving the mind as it is the unplanned impedimenta of thought and rationalization, accidents of body chemistry or neurological misfirings. The Pace of the Stars is a system. It is a method that is taught and learned and put into practice, and has been for over a thousand years. Inasmuch as there can be a science of coincidence or synchronicity—a plan for acquiring knowledge outside the realm of rote memorization that was such a hallmark of the mandarin class in China—Daoist meditation practices provided that framework. By turning the focus of the mind inward and withdrawing as much as possible from the external world and its sensory messages (messages that are by necessity reduced to those that the eye and ear can sense, which means levels of vibration that occupy but a small fragment of the entire spectrum of light and sound) the Daoist adept was able to see connections between phenomena that otherwise would have gone unnoticed. The Neolithic Chinese shamans—the *wu*—with their ecstatic flights, their trances and crazed hopping dances, became the kings and emperors of the oldest country on Earth. They formed the bridge between This World and the Otherworld that empowered secular as well as sacred rule. It would be a serious mistake to suppose that this rulership was the result of sophisticated con-men pulling the wool over the eyes of a credulous populace. That would come later. In the beginning, survival was at stake: survival of the family, then the community, then the kingdom. The role of the shaman was critical to the smooth functioning of a society that relied so much on the vagaries of nature: of the hunt, of plants, of the seasons. The shaman understood nature in a way that others did not. And this understanding, according to their own accounts, was derived from passage to the heavens.

Again and again we are forced to confront the single most overwhelming fact of human experience: that there is a solid connection between humanity and the stars. It is not one of mere longing towards an off-world paradise that we have somehow imagined must exist (all evidence to the contrary), but of a memory buried so deeply in human consciousness that every time anything unexpected occurs, anything out of the ordinary, whether lights in the sky or mutilated cattle or dreams so realistic that they are more real than waking life, we associate it with space, with extraterrestrial beings, with gods and angels and demons and jinn.

What the Daoist magicians did—and the Indian Tantrikas, the European alchemists, the sorcerers, the witches, and all the exotic practitioners of antinomian spirituality—was to accept the extraterrestrial hypothesis as a given and work from there. "How do we do this?" they asked. "How can we travel to the stars?"

They knew the answer was within themselves, at some nexus between body and not-body. They instinctively knew—millennia before the genetic code was discovered—that they carried within themselves the answer to this existential dilemma. They would write of duality and non-duality, of the Dao, of transcending yin and yang which means transcending Earth and Heaven, both. They would write of the origins of Creation, the origins of humanity, as something that was derived from the stars. They dug elaborate chambers for their dead and for their gods, Chinese versions of the Tomb of Osiris, aligned to the stars. They took the Egyptian model one step further: building on the idea that the dead travel to the heavens, the Chinese sages and shamans asked the obvious question: why not the living, too?

And so began the development of a technology for space travel predicated on the ability of human consciousness to do the heavy lifting. It had to be consciousness. Consciousness was what made us human; it's what differentiates us from everything else on the planet Earth. Therefore, consciousness was the connection between human beings and their celestial origins. It would be consciousness that would provide the vehicle for making the ascent.

In order to control consciousness one has to begin with controlling sensory input. One has to reduce the noise as much as possible so that only one signal can get through. Darkness, silence, restricted movement, attention to breathing, fasting, abstention from sexual or even social contact . . . the ancient cosmonauts were imitating the conditions of twenty-first century spaceflight. The right time had to be selected, the right atmospheric conditions, the right orientation. The body was not the vehicle; instead, the body was the launch pad. What would do the traveling would be something else.

The Egyptians had a very complex breakdown of the human being as we saw in the previous chapter. *Ka, ba, akh, ren, sheut, ib,* etc., were conceived as different aspects of the human being from the body to the name, soul, spirit, combined essence of soul and spirit, etc. While these aspects were all present (except, perhaps, for the *akh*) during a person's life and especially during the life of the pharaoh, they only became very important after death when finding the right balance of all of these aspects would ensure immortality. At least, this is what we assume. Since the bulk of the religious hieroglyphic inscriptions from the Old Kingdom to the New Kingdom appear to be concerned with death and mummification, there is no way to know if the priesthood actually entertained other ideas and if they, in fact, did practice some form of traveling to the stars while still alive. After all, is it reasonable to ask where the information in the

Pyramid Texts and other funerary documents came from if not from a living person(s) who had already made the trip?

If, as Robert Temple has suggested, the origin of the pyramid builders is to be found not in Egypt but in northwest Africa among the megalithic culture that erected stone circles in Morocco,[22] then it is possible that a form of shamanism was practiced among those who eventually made it to Egypt and created—or contributed to—that country's ancient civilization. Conversely, if researchers such as Robert Schoch and Stephen Oppenheimer are correct, then knowledge of astronomy, sacred geometry and astrally oriented architecture was brought to the rest of the world by the refugees from sunken Sundaland, a science which may have included shamanistic celestial flight. Either way, it seems reasonable to assume that the specificity of detail concerning Otherworld journeys was derived, at least in part, from voyagers who had already made the trip: either out of the body, or in the body. While ancient astronaut theorists usually are ready to ascribe alien visitations as the source for many religious and esoteric practices among the ancients, the reverse—that human beings made the trip themselves—is usually not given much attention, even though modern cases of "alien abduction" proliferate in the literature. What is the difference, then, between an alien abductee and a Daoist treading on the stars? The difference is one of context rather than distance. At some point one of our Cargo Cultists hopped a flight— stowed away, maybe—and made the trip.

The basic requirements of celestial flight remain remarkably similar from culture to culture. Sensory manipulation of some kind is essential: either through drumming and chanting—the hypnotic, rhythmic repetition being the common denominator—or through intense meditation incorporating sensory withdrawal, or even through the use of hallucinogenic substances, the method involves dislocating one's sense of reality, of This World, and turning the focus elsewhere. The mind seeks a place to fix its attention, and when it is deprived of a focus among the sensory inputs of waking reality it will find a target farther away. When it does that it will begin processing the data that comes in and will try to incorporate it within what it already knows or has already experienced.

Those who have already gone on before—have entered the shamanic trance or have had their minds altered in some way, their body chemistry changed, their neurological system tweaked—will report, in the vocabulary available, what has transpired and those who have not yet made the trip will try to imitate the process as they understand it.

We can use the metaphor of a flight simulator in order to understand the shamanic trance that is at the root of both the Chinese and the Egyptian mysteries. The rituals may be intended to instruct, to train, rather than to accomplish the seemingly impossible: immortality, survival after death, celestial flight, etc. If so, then what is the purpose behind the training? By duplicating as much as possible what the experiencer has seen and learned, others may follow.

The fact that we have UFO religions today is evidence of this phenomenon in action. The experiencer becomes a shaman, a priest or priestess of the Alien Gods. It is not

a new phenomenon, as we have been at pains to demonstrate. It is an old one, as old as humanity's first contact
with things not of this world, and as such actually demonstrates the validity of our argument that—at heart—*all*
religion is UFO religion. Once we come to grips with that,
so much else becomes easier to understand.

▼   ▼   ▼

We tend to live a bifurcated existence, split between who we
are as human beings and the external world, the real world.
We are required to think of our interior lives as not-real:
our dreams, our memories, our imagination, our emotions.
What is real is what can be observed and measured. Life is
real, and so is death. Virtually everything else is subject to
interpretation. Many of us have been accused at one time
or another of not having a grip on reality; the assumption
being that reality is a solid, stable object that is subject to
slipping through the metaphorical fingers of our minds. We
are told not to trust our thoughts or our emotions. In other
words, we live a life in constant struggle and engagement
with this absolute substance we call reality. We engage with
reality through our senses, but we are told that our senses
reveal to us a subjective version of reality. Eyewitness testimony, we are told, is notoriously unreliable.

We therefore invent machines that are more reliable
in measuring and weighing the real and we rely upon
the machines and ignore our own senses. This means we
engage with machines more than with Nature, or with

other human beings, as the machines are always "right" and the rest of creation is not wrong exactly, just unknowable without the machines.

We dream, and ignore the content of those dreams. We explain away the phenomenon of dreaming using all sorts of mechanistic rationalizations that focus on brain chemistry, which can be measured, or on psychology which is just another form of narrative, another way of dreaming while awake.

Dreams show us scenes of life, but they are not real. Dreams are irrational. They don't make sense when we are awake, although they make excellent sense while we are asleep and in the grip of the dreams. Since, when we are asleep, we no longer are engaging consciously with the world around us—with reality—whatever we experience must be not-real. Irrelevant. A kind of temporary psychotic state. Religion, therefore, is in the same category because of the stories in every religious text that describe things that are not real or only as real as dreams: gods, angels, demons, spirits, immortal beings, life after death, transcendence, immanence, resurrection, apocalypse.

This litany of the irrational and the non-existent is often extended to include: space aliens, flying saucers, political conspiracies, secret societies, satanic cults, ancient astronauts. Whatever the machines we invent cannot measure or record.

The practice of shamanism—and of mystical practices generally—is the one arena where consciousness is all that matters, where there are no mediating machines except the ones that already exist in the human brain and, by extension, the human mind. As we will see in more detail in Book Three, the machinery of the human nervous system is so intricate and well-designed that it still cannot be duplicated by computers or artificial intelligence systems. Consciousness studies may provide the key to understanding the universe in ways that leapfrog over existing hard-wired technology. Various means are proposed for understanding how consciousness "works," from microtubules in the brain to electrical fields around the body. The ritual specialists and shamans were never that concerned with the nuts and bolts of the phenomenon, just with the methodology. At some point—perhaps as recently as twenty thousand years ago—human beings formed images in their brains of things that did not yet exist: tools, calendars, sowing and reaping, astronomy, architecture. They made the real world adapt to their imaginary world, the world of their visions and dreams. And once they started, they could not stop.

Running parallel to all this activity was the trance of the shaman. That did not change, except to become more sophisticated in terms of complex diagrams of ascent and descent, of the names of gates and guardians and Underworld spirits. The basic technology, however, remained the same. As toolmaking became more advanced, the shamanic trance became relegated to the backwater of human experience: to superstition, credulity, and deception. Now, however, it has begun a kind of renaissance, attracting attention from academics and New Age seekers alike.

But in ancient China, it was called *wu*.

▼   ▼   ▼

The character for *wu* or shaman consists of two human figures, one on either side of a central figure that is usually taken to mean "work" (i.e., *gong*). However, recourse to other versions of the character indicate that the *gong* figure may represent something else, possibly a central pole or axis mundi. The character is believed to have evolved from a much simpler figure, the "cross potent": an equal-armed cross similar in design to the Templar cross which is said to represent a magician. In fact, some etymologies of the Chinese words for *shaman* are linked to those for *magician* in Indo-European languages. It is a little startling to find the Templar cross in use on Chinese bronzeware dating to the mid-second millennium BCE, and intended as a reference to magic and shamanism!

The ancient bronzeware character for *wu* or shaman.

Zhou Dynasty Seal Script character for *wu*. Note the two individuals on either side of a central pole or axis mundi.

Thus the Chinese character or ideogram for *shaman* is illustrative of shamanic practices, showing the importance of the central pole leading to the North and the dancing shamans around it. Research being undertaken by scholars at this time strongly supports the view that Chinese shamanism was not concerned only with trance states, divination, and healing but was also involved in "actively creating a connection between heaven and earth"[23] as early as the Neolithic era. The design and orientation of the structures found at Niuheliang reflect this connection and also support the idea that the shamans were themselves the political hierarchy of ancient Chinese society. There would be no point in asserting social control in the name of the stars unless it was understood at that early age that authority came from the Other World (or Doubleworld) and not from human leaders per se, indicating (or at least suggesting) some type of paleocontact with the Other World.

The shaman was the human link between the Earth and the Heavens at a time when there was no perceived difference between the state and religion. Entry into the exalted realm of the stars was accomplished by means of altered mental states, consciously and carefully controlled. It was not simply a case of "turn on, tune in, and drop out" as the Sixties mantra had it. Rather, it was a deliberate manipulation of the neurological system and a constructed disorientation of the senses so that one lost one's conscious moorings in This World only to be set adrift in the Other. That alignment to celestial sky marks was essential—a kind of dead reckoning of the soul—indicates that the travel took place in both space and time according to a terrestrial-celestial arrangement that was decided, calculated, in advance.

Indeed, it is known among scholars of ancient Greek that the word from which we derive the English "mathematics" is *mathematikos*, which means "astrologer." Even the word *mathesis*, which is usually translated as "learning," also is used to mean "astrology." There is thus a very old tradition that equates mathematics and learning in general with knowledge of the stars. Of the two forms of divination known to the ancient Chinese, one—the oracle bones— was considered the "rational" version as opposed to the shamanic trance which was the more chaotic form (and still used today by the Dalai Lama and his State Oracle, a man who undergoes an intense trance and once "possessed" speaks in a language intelligible only to the Dalai Lama). The oracle bones are our earliest evidence for a Chinese writing system and it was developed to tell the future. Thus writing had as much a mystical, occult character in ancient China as did hieroglyphics in ancient Egypt. There may be a connection between writing—which is communication at a distance in both time and space—and the Phenomenon.

In order for the celestial flight of the shaman to be successful the square table of the Earth had to mesh its gears with those of the round table of the Heavens. Calculations of the rising and setting of astral phenomena were central to Chinese ritual and architectural alignments. This shows us that there is a very pragmatic purpose behind the "harmony" and "balance" that are such important characteristics of Asian philosophy. Without that harmony and balance the celestial flight cannot be accomplished and the rocket explodes on the launching pad or, at best, sets the astronaut adrift in the cosmic void with no way to return to Earth. Yet, we hasten to emphasize once again that there is no obvious evolutionary purpose to any of this and yet similar systems can be found all over the world from Neolithic times to the present. Our earliest writings and artwork were of a religious, magical, or supernatural nature and demonstrate contact with forces, races, or entities that originated outside the accepted evolutionary context of humans evolving from other primates. This is the Phenomenon in its earliest known manifestation and it jump-started human civilization.

The Chinese context places an emphasis on the role of the human mind and especially the human nervous system as the machine to be used to effect this contact. This is a deliberate contact, not a passive "experiencing" but a carefully planned technique for ascending to the stars. We do not make a space for this type of technology or "machine" in our mainstream academic dialogues because anything smacking of a work of the imagination is not held to the same standards of realism as the externally crafted tools of the engineer and the scientist. Yet the "sekret machine" of the human brain is a main contact point with the sekret machines that continue to populate our skies. The training of the human mind in the celestial ascent technologies of the shamans, the *wu*, the Jewish mystics and most probably the Egyptian priests will prove to be an essential element of the type of space travel that will take us beyond the solar system and to the stars. The Chinese systems clarify and expand upon the Egyptian systems: it is a dialogue between shamans that has as its heart the next step in our evolutionary path, which includes contact—deliberate, cautious, and consciously controlled—with the gods of our alien genesis.

This is Doumu, the Chinese goddess of the North Star and the Dipper. She is the Mother Goddess of humanity, along with her consort Shangdi, the "Father of the Plough" (i.e., Big Dipper.).

Another divine couple, FuXi (right) and NuWa are considered to be the creators of humanity. They have serpent bodies and human heads, and the serpents are coiled around each other. Refer to the previous chapter illustrations with depictions of Ningizzida, for instance, or the Gnostic icons, all of which show serpents coiled around each other or as bodies for divine beings. In this case, this is an illustration unearthed at Xinjiang Province (the far western province of China), and shows FuXi holding an instrument which is strangely similar to the Egyptian adze used in the Opening of the Mouth ceremony. His consort, NuWa, is holding what appear to be compasses. Thus, the instruments have resonance with Masonic imagery, as well. The beings are surrounded by heavenly constellations. The one just above the head of FuXi is the Big Dipper, or "Celestial Chariot."

KALA CHAKRA/CHAKRA VARTIN

Shambhala . . . is sometimes described as like a pure land, a place beyond the reach of ordinary travel . . . initiation is said to establish predispositions for rebirth in Shambhala not only for the sake of maintaining practice of the Kālachakra system but also for being under the care and protection of the Kulika Rudra With A Wheel when the great war comes.[1]

– Jeffrey Hopkins

The psychological experience that is associated with the UFO consists in the vision of the *rotundum*, the symbol of wholeness and the archetype that expresses itself in mandala form. Mandalas, as we know, usually appear in situations of psychic confusion and perplexity.[2]

– C.G. Jung

WHILE THE CHINESE SYSTEMS WE LOOKED AT IN THE previous chapter demonstrated an extension of the Egyptian journey to the Underworld as an interior, spiritual process the Indian concepts take that one step further. While the Daoist adepts understood that they had to internalize the stars of the Dipper in order to align themselves properly with that asterism, the Indian sages identified each of the stars with specific areas of the body—called *cakras* or "wheels"—along a central Pole and discovered neurological and psycho-biological analogues that enabled them to accomplish the same feat. For the Indian adepts, as for the

Chinese, each of the seven stars of the Dipper had associated deities, passwords (mantras), and other attributes that helped train the mind of the practitioner and prepare him or her for the celestial ascent. Also, the Indian system of sacred architecture reflects the same concepts as those of the Chinese and the Egyptian with its emphasis on the North and the polestar in orienting temples. Immortality and celestial flight are also traditional aspects of Indian mysticism, as we will see.

Ancient astronaut theorists have cited something called the *Vymanika Shastra* as a text revealing "flying saucer" or *vimana* technology as experienced in Vedic times. However, it has been demonstrated[3] that the *Vymanika Shastra* is a twentieth-century "channeled" work and as such does not reflect genuine Vedic sources. While not a hoax, its value in terms of this discussion is not very great. For that reason, we will not spend any time discussing its claims as that would get in the way of the genuine Vedic and Tantric material, which is far more fascinating and revealing than the *Vymanika Shastra*. Like many ancient scriptures, the Vedas record the flights of gods and other beings, often in fantastic vessels. This fact alone is suggestive of some kind of experience or contact with human or quasi-human beings that were able to travel through the air, an experience that became associated with various legendary or historic events in India's past.

▼   ▼   ▼

The period in which the Vedas were composed usually is ascribed to the middle of the second millennium BCE or about 1500 BCE to 500 BCE. Thus the Vedas were being transmitted orally in India about the time of the Egyptian New Kingdom and the Chinese Shang Dynasty. It is believed that the information contained in the Vedas is considerably older than its earliest written forms and probably reflects a Bronze Age culture that developed out of the Indus Valley civilization when the latter collapsed about 1900 BCE, leading to a migration of the peoples known as Indo-Aryans to the Indian sub-continent.[4] There are many similarities between the religious concepts detailed in the Vedas and those of the ancient Iranian religion, which has led to a close identification of the two cultures. While most of the Avestas (the sacred texts of the Zoroastrian religion) have been lost due to the Islamization of Iran (Persia) there are surviving fragments as well as the persistence of some ancient Iranian concepts and terminology among the so-called "Kafiri" of Nuristan (a non-Abrahamic culture in Afghanistan that was converted to Islam only in the late-nineteenth century, and which was immortalized in the Kipling novella, *The Man Who Would Be King*; the term "Kafiri" is a pejorative Arabic word meaning "infidel") and the Kalash people of Pakistan (across the border from Nuristan).

As we examine some of the Indian concepts related to our study of the Phenomenon, we have an opportunity to compare Zoroastrian terms, concepts and iconography to the Vedic examples. Zoroaster is said to have lived in eastern Iran about 1500 BCE, thus making him a contemporary of the Vedic period in India as well as the New Kingdom in Egypt. The Avestan language, in which the Zoroastrian scriptures are composed, dates from about the same period and has some similarities to Sanskrit, the language in which the Vedas are composed.

Perhaps the most famous icon of the Zoroastrian religion (and of the nation of Iran) is the *faravahar*. This is an image of a bearded man sitting in a circle that contains spokes like a wheel (or rays, like a star) and which itself bears two great wings. It is said to be a version of the *fravashi*: a kind of guardian spirit.

THE AHURA-MAZDÂ OF THE BAS-RELIEFS OF PERSEPOLIS.[1]

The above image is of the Zoroastrian deity Ahura-Mazda. In this version—taken from a work by the Egyptologist Maspero—the deity is standing inside a winged circle. In an earlier bas-relief we see:

This is the same concept, but in this case of the Assyrian god Ashur. The rays or spokes clearly can be seen in the circle or wheel behind him. The two long curls that descend from the circle of Ahura-Mazda seen previously are here on top of the circle. Thus we are looking at a persistent motif that probably had its origins in Mesopotamia among the Babylonians and Assyrians and which later made its way to Persia. It was a familiar icon of supreme divinity: a human or human-like figure flying above the Earth in a winged, rayed globe. In the case of Ashur, we have the god holding a bow, an instrument of war. In later images the being carries a ring.

The following example is from the Nimrud excavation by A. H. Layard and currently on view at the British Museum:

This stele dates from the mid-ninth century BCE. In this case, the god is shown holding a ring in his left hand. It is obvious that he is flying: he is shown in the air above the standing figures and if we apply a certain perspective to the art—assuming the god is at least as tall as the figures on the Earth below—we can propose that the god is far above the standing figures and perhaps deeper in the background.

Various explanations have been advanced for this design. Depending on the source it may be identified as a solar symbol—the circular sun with its rays, the wings signifying its travel across the sky—or as a symbol for divinity itself. The human figure in the disk may indicate the god of the sun, for instance. (In ancient Sumerian iconography the symbol for divinity is an eight-rayed wheel or star.)

It is believed that the original form of this symbol was used to designate Ashur, the Assyrian god, who may not have been a solar deity. In fact, the origins of Ashur are not known with any degree of certainty. He was considered among the Assyrians to be the "great god" or the ruler of the Assyrian pantheon. At some point the symbol for Ashur became the symbol for Ahura-Mazda, the supreme god of the Zoroastrians whose name may have been derived from Ashur.

Ahura is an Avestan word which is derived from the Indo-Iranian word *asura*, which is then identified with the Sanskrit *asura*. This is important, for it points to a specific conflict between classes of divine beings who are at war with each other. In the Indo-Iranian context, the asuras are angelic beings who fight the *devas* (from which we get both the English words *divine* and *devil*). In India, however, the devas are benevolent beings and the asuras are malevolent.

The common denominator is conflict, and especially conflict between two non-terrestrial camps. This conflict is experienced by human beings in some way, and even though stories about this conflict make their way into the sacred texts of different cultures, there is no immediately obvious reason why they should be included since they do not have a direct impact on human life or development. Indeed, the strange way in which the asura-deva conflict is reversed—in Iran, the asuras generally are benevolent, in India they are malevolent—may indicate that human beings have taken sides in the conflict. In fact, some scholars point out that the word *ahura* in the Zoroastrian texts sometimes refers to human beings or beings with human characteristics who are military commanders or otherwise have exceptional abilities.[5] The term is used to refer to human lords as well as to "other" lords of the same lineage as Ahura-Mazda: the Supreme God of the Zoroastrians. There thus seems to be a connection between divine and human rulers with the human rulers partaking of some sort of essence of their divine counterparts.

The Zoroastrian usage of *deva* indicates a category of gods that are "rejected," and will later become known as demons. In India, the devas are gods or, as Coomaraswamy has suggested, "Angels" versus the "Titans" of the *asuras*,[6] beings that are nevertheless "consubstantial." This last is an important point for it is consistent with other views, such as those in the Abrahamic tradition, which state that the devils are fallen angels. There is thus a degree of consubstantiality between the angels and the devils of Jewish and Christian lore, which reflects the same "kinship" between the devas and the asuras whether in an Iranian or Indian context. This general agreement transcends geographical boundaries and religious cultures: there was a war in heaven, a conflict between the gods, that eventually had ramifications here on Earth as humans were enlisted to support one side or another. This conflict is highlighted in the doctrines of Manichaeism: a third century CE religious movement that originated in the Persian empire with the prophet Mani (216-276 CE), who was born in what is now Iraq. Mani's religion was heavily influenced by Jewish, Christian and Gnostic ideas concerning the creation of the world and the struggle between forces of light and darkness. He composed an elaborate cosmological system with different levels and stages of Creation including a scheme of cosmic wheels which are set in motion by the gods. At one point Evil (in the form of male and female "archonts" or "sons of Darkness") is transformed into a giant sea monster who is slain by the "Light Adamas," the dragon-slayer, and who is then stretched out from east to west and the north to form the cosmos (in an obvious reference to Mesopotamian Creation literature involving Tiamat and Marduk).

The war between the powers of Light and Darkness began before the cosmos was created, before the creation of human beings on Earth. This was a conflict that had its origins before what we would call the Big Bang. According to this doctrine the powers of Darkness were contained in the South while the powers of Light occupied the other three quadrants of North, East and West. Obviously, if there was no cosmos as yet and therefore no Sun and Earth, cardinal directions did not yet exist, but this narrative should be interpreted allegorically. This dualistic concept of primordial reality is reflected in many other cultures, such as the yin and yang of Daoism. In Manichaeism, Light is trapped in Darkness and the cosmos was crafted as a large wheel which would separate the Light from the Darkness and send the Light back to the realm of the gods. In a sense, then, this was seen as a refining process and as such is cognate with ideas in alchemy which see gold—the perfect metal—as trapped in baser materials, requiring only the dedication and perseverance of the alchemist to separate it out.

Many of the original Manichaean scriptures have been lost or destroyed, some at the insistence of Zoroastrian and Mazdaean officials of the Persian empire who saw the new religion as a threat to their hegemony. Regardless, Mani's influence spread to India—where it is said he sojourned for awhile, healing the sick—and to China, where a Manichaean temple is said still to exist. Manichaean religion and doctrines spread as far west as Africa and as far north as Europe and England. At one point it could be considered a world religion as it stretched from China to India, Europe and Africa, in its heyday, and lasted for more than a thousand years. It influenced such later Christian heresies as Bogomilism and Catharism, both of which are well-known for their relationship to the Templar legends and the iconic Holy Grail.

One of Mani's major influences was the *Book of Enoch*, and its associated *Book of Giants*, copies of which were discovered at Qumran in 1947. The *Book of Giants* is especially interesting as it is based on the story of the Nephilim: the controversial term found in the biblical Book of Genesis and its relation to the story of the Great Flood. Made popular by ancient astronaut theorist Zecharia Sitchin, and discussed here in an earlier chapter, *Nephilim* has been translated variously as "fallen ones" and as "giants." Once again we are in strange territory: the persistent idea that there was a kind of genetic abnormality—there is even evidence that the term understood by Mani to mean "giants" and "monsters" was also used to mean "abortions"[7]—on the Earth that was somehow linked to affairs in the Heavens and which resulted in the Great Flood, is found again in the writings of Mani and especially his version of the *Book of Giants*. It is a familiar theme in alien experiencer circles that hybrids exist on Earth who have genetic donations from both human and extraterrestrial sources. One can suggest that these hybrids would be equivalent to the giants, monsters, and even the "fallen ones" of biblical and Manichaean doctrines. What is interesting is the fact that it is somehow necessary to talk about giants, monsters and fallen ones at all in this context. Be that as it may, after the appearance of these genetic abnormalities there is the scenario of the building of a vessel and of the escape of a small number of human beings—along with collections of animals and plants—from the ravages of a deluge. We saw this not only in the Mesopotamian context but also in the Aztec context: different races of beings, in conflict with each other, and the survival of one race after a flood. The biblical version insists on a supernatural or extraterrestrial context for the conflict that led to the abnormalities and from there to the flood and the survival of a small fraction of humanity (presumably those not affected by the genetic misfiring).

The Vedic text *Satapatha Brahmana*, composed about the mid-first millennium BCE, also recounts a story concerning a flood, being the earliest known Indian flood legend.[8] In this case it is the story of Manu and Matsya. Manu is a human being, and he is warned by a fishlike creature—Matsya—of an upcoming deluge. He prepared a ship at the appointed time, boarded it, and the floods came. When they subsided, the ship beached on the summit of a northern mountain (as in the biblical version). Manu descended from the ship and began making sacrifices in order to produce offspring, for he was childless and evidently without any human companions.[9] A woman was produced from the offerings of the sacrifices who called herself the daughter of Manu, but with whom he produces the rest of the human race. Other, later, versions of this story are present in the *Mahabharata* and the *Puranas*, with some differences and further elaborations.

The reason behind this arcane history lesson is to introduce some of the Indian concepts concerning alien contact and to show that there is some commonality of themes running through the sacred texts of the world's great religions. What seems to be myth and imaginative storytelling to modern eyes had a different reputation in ancient times. These were the records of events that had taken place for which, as we have said repeatedly, a scientific vocabulary was not available. Eventually, some core aspects of these records became the basis for practices and beliefs concerning traffic between the stars and the Earth, through the medium of non-terrestrial, non-human agencies. By understanding this, we can better approach the modern experience of alien contact and alien abduction by seeing them as aspects of a Phenomenon taking place along a continuum from ancient times to the more familiar and contemporary military, government and personal documented accounts.

As vocabulary became more precise, more specifically designed around scientific terminology and concepts that were based on observations and measurements of the external world, human language changed accordingly. Scientific vocabulary became the lingua franca of commerce and technology and contributed to a widening gulf between religious and spiritual concepts on one side and the language of the "real world" on the other. As language underwent this separation, religious and spiritual vocabulary became devalued over time and this has contributed to a kind of atrophy of the human understanding of the role of psychological and emotional functions of consciousness. Thus has led to the current tendency to regard the language and art of pre-modern societies as swamps of superstition and credulity. At the same time, language associated with esoteric practices became increasingly encoded as writing and other forms of communication began to dominate the channels of instruction, knowledge and education. The dominance of digital media—due to the widespread adoption of electronic forms of knowledge transmission—has led to a suppression of more analogue forms or at best the relegation of them to the arts, music, etc., where they do not pose a threat to the technology priesthood. The results of this split can be seen in everything from economic inequalities among those who do not have access to digital sources of information or expression to the rise of religious fundamentalism among those who sense the impotence of technology to satisfy deeper yearnings. The unspoken threat in the modern world—the gorilla (guerrilla?) in the room—is the fact that technology has replaced religion as the opiate of the people, and the backlash of religious fundamentalists is a predictable reaction to this realization. Ironically, both sides have become seduced by machines.

Sator Arepo Tenet Opera Rotas[10]

> The *ratana-cakka*, the ideal wheel, is described as the divine wheel that appears to one who is destined to be a *cakkavatti-rājā*, a universal monarch.[11]

The symbol of the wheel will take us deeper into some of the more esoteric aspects of Asian thought. The primary influence of the Dalai Lama on non-Tibetan societies, for instance, has been the promulgation of the *Kalacakra Tantra* and its associated initiations. The term *kalacakra* means "wheel of time" and refers to a complex of ideas expressed both in language and in iconography. For whatever reason, the symbol of the wheel has become a central image in Indian and Tibetan religion and esoteric practice. The cakra itself is inescapable in everything from the everyday practice of yoga to the highest and most secret initiations into various Tantric sects. The wheel appears on sacred drawings, carvings, bas-reliefs, in a variety of contexts but always as a signifier of power, rulership, and illumination. And, as the above quotation reminds us, the appearance of that wheel indicates that the observer has attained a special status. Like the shamans we discussed in the previous chapter, the Indian rulers were those who had made contact with the Other World. In China, heaven is a circle; in India, a wheel.

The Indian version of the wheel is described as flying through the sky:

> It is stated that a king having perfected the ten virtues of a universal monarch observes the eight precepts on a full-moon day and then retires to the top-most floor of his mansion, when the divine wheel rises from the eastern Ocean and comes through the sky like a second full moon. It circumambulates the mansion . . . On the command of the monarch, the great wheel starts on its mission and the conquest of the world begins.[12]

This is obviously not a reference to a solar disk or some other planetary or stellar object but a vehicle of some kind that only appears when the king has undergone a period of purification and right action. At that point the wheel appears to the king, who sends it forth on its mission of conquest. Thus, the wheel is both a vehicle and a kind of weapon.

This description comes from the Pali Canon, the basic texts of Theravāda Buddhism and the earliest known Buddhist scriptures. The wheel is an important Buddhist icon and as recognizable as a symbol of Buddhism as the cross is of Christianity or the crescent of Islam. Usually it represents the Wheel of Dharma, the Wheel of the Law, but it can also mean the Wheel of Life (and many other associations as well). The ancient term for an ideal ruler (Buddhist or Hindu) is *cakravartin*: "he who turns the wheel." While the instruction concerning the king and the divine wheel may be at least partly fantastical and allegorical, the fact that a vision of a wheel the size of the full moon circling around the king's palace and then going off to conquer the world is accepted as a valid image indicates that ancient Indians saw similar lights in the sky and associated them with kingship and with the power to control secular affairs, power that later came to be interpreted as the Law, or Dharma in the Buddhist context.

Evidence such as this is more important than channeled *vimana* documents of impossible vehicles. To be sure, it is subtler and not as striking as the weird designs of starships in the *Vymanika Shastra*, but textual information such as this is relatively unimpeachable and at least as relevant as the vision of Ezekiel or the Ahura-Mazda symbols. Archaeologists have suggested that these wheel symbols represent the solar disk, or the solar disk with rays radiating from it, and other interpretations which assume a near obsession with the sun among ancient peoples. That archaeological sites are demonstrably oriented towards solar and lunar positions as well as to the Dipper may be taken as further evidence of this obsession.

How, then, would the Phenomenon be interpreted, either textually or in iconographic form? It would be mentioned in such a way that no association with the sun or moon could be construed. In terms of a text, the sun or moon might be mentioned as background to the larger story (as in our example above, in which the full moon is mentioned in a deliberate way as being different from the "divine wheel" which then behaves in a way that neither the sun nor the moon ever behaves). In terms of art, there would be details suggesting that—even though there was superficial resemblance to the sun or moon—it was not to be confused with either. If an ancient observer were to see a flying disk, for instance, how would it be portrayed in a manner that makes a distinction between it and other celestial phenomena?

One would begin with a circle, the basic design and the common denominator of sun, moon, and flying disk. Then one might add identifiers that suggest a unique quality of the disk, either in terms of its movement or its appearance. In the case of the symbol of Ashur and of Ahura-Mazda, there is a human or humanoid figure in the center of the disk. The figure holds an object or objects that suggest function, such as the bow of Ashur or the ring of Ahura-Mazda: clearly not objects that would be associated with either the sun or the moon, or at least not without a lot of explanation.

There are wings extending from the disk, both to the sides and below. Some archaeologists or religious scholars have suggested that these are not wings but are intended to represent the rays of the sun. If this is so, then why do the rays extend only in three directions and not in all directions? Why are there no rays extending from the top of the disk? It seems clear that these are intended to be wings, and specifically the wings of a bird in flight. In some examples, what appear to be rays can be seen within the circumference of the disk but not extending outward into space. These appear to indicate the spokes of a wheel rather than the rays of the sun. This suggests that what we are looking at it is a *device*, not a common celestial object.

The sun did not suddenly appear over the skies of Mesopotamia one day, causing people to wonder at it and develop all sorts of mystical associations with it. People were born, grew up, lived their lives, and died under the sun's rays since the beginning of life itself. For hundreds of thousands of years, *Homo sapiens* was aware of the sun. Of course, as intelligence developed and the rudiments of civilization were created—including speculation about the role of human beings in the cosmos—the sun, moon and stars became objects of veneration and calculation. Narratives were created to explain or enlarge upon observed celestial phenomena. There came a time, however, when celestial phenomena were not predictable or calculable: comets, meteors and other natural events were observed, recorded, and wondered about. Then, in the midst of all of that a different kind of phenomenon took place, one that inspired the association of specific human and functional attributes to something sighted in the sky. A disk with a man in it, or something like a man, with the "likeness of a man" as we read in Ezekiel. Something circular, like a chariot wheel, spinning. Flying. Something not natural, but something not manmade, either. Something alternatively terrifying and awe-inspiring.

The closest approximation of this flying object was its similarity to a wheel, but it was a perfect wheel: an ideal wheel, one that spins without friction, one that cannot be stopped. One that spins effortlessly. One that stays aloft.

The Wheel of the Law. The Wheel of Life. The Wheel of Time.

Kalacakra

The central text of the type of Buddhism most identified with Tibetan Buddhism is the *Kalacakra Tantra*. Since his investiture as the fourteenth Dalai Lama, Tenzin Gyatso has performed numerous Kalacakra initiations around the world and in some cases of hundreds of thousands of people at a time. The wheel is the central symbol of this initiation.

The Wheel of Time is a *mandala,* a sacred diagram that is composed of a specific number of elements in appropriate positions and colors. The Swiss psychologist, C. G. Jung, identified the mandala as a device indicating wholeness; when his patients had dreams of mandalas it was evidence that they were approaching the stage of individuation, a psychological state in which the various aspects of a personality were coming together to form a harmonious whole. The appearance of the UFO, according to Jung, was an indication that humanity in general—in a state of confusion and perhaps fear—was yearning towards that state of harmony and unity and projecting that yearning towards the sky.

The mandala imposes order upon chaos. Its balanced elements—usually in multiples of four—indicate that everything is incorporated into the state of order without favor of one element over another. Such objectivity is normally beyond the capacity of human beings who cling to one image or sensation over others. In fact, the requirements of many initiatory systems make serious demands upon the psychological, mental and emotional resources of most individuals. The suppression of physical and emotional desires that is the hallmark of Tantric initiations, for instance, as it is of monastic life in the West, may be related to a desire or need to emulate the perceived state of non-terrestrial contacts. In other words, physical and emotional states that are not natural to human beings may reflect the natural state of non-human beings. This is akin to the Cargo Cult phenomenon with which we began this study: the impulse to imitate the perceived behavior of colonizing powers if that behavior is understood to be related to superhuman or supernatural abilities. This is especially relevant if the only two possible reactions to the presence of a colonizer is resistance or emulation.

If the non-terrestrial power is not intent on colonizing and its actions are not perceived as threatening or repressive, then its behavior may be even more attractive as an attribute worthy of imitation (or, as in the current global situation regarding UFOs, merely ignored). The distance created between the alien power and its human observers makes a space for the projection of ideas about that power. Human beings tend to understand the world in terms with which they are already familiar: love, fear, hatred, aggression, greed, piety, profit and loss, life and death, etc. This is why the question always remains: Why do aliens not land on the White House lawn? From the perspective of a putative alien force, why would they? We attribute human strategies to beings that do not share our evolutionary path on this planet, which is certainly a mistake.

The only people on Earth who might begin to understand an "alien" mentality would be those who distance themselves as much as possible from the realm of "human" action, the better to see alternative perspectives. This does not mean that these individuals actually do perceive or understand alien intentions and thought processes, but the distance created allows room for the introduction of new and more promising ways of comprehending other possibilities through a medium that even may be nonverbal.

Human language presupposes human experience related to this planet. The famous plate affixed to the surface of the *Voyager 1* space probe showing a nude human male and nude human female waving (a gesture that—to an alien consciousness—may have none of the associations we attribute to it even if it is perceived as a gesture at all) along with other data that we assume would be understood by an alien being is a case in point. It would presumably be understood by other human beings, but the presumption ends there.

In such practices as Tibetan Buddhism or Tantra—and many others—we have systems of language, behavior, and intellection that are counterintuitive. The language of the Tantras, for instance, is known variously as "twilight language" and "intentional language." It is a system of coded transmission that relies less on normal grammar and vocabulary and more on a network of associations and multiple meanings in order to transcend language as we understand it. That does not mean, of course, that an alien intelligence "speaks" this language; it does, however, provide an alternate approach to the idea of language and how it may be used as a deliberate symbol system or even transcended in order to reach a nonverbal state of communication.

Experiencers often have described communication with alien contacts as nonverbal and telepathic: as the transmission of images, or as sounds heard within the brain rather than as external stimuli. Rituals such as those of the *Kalacakra Tantra* employ images (such as the mandala) in an active manner, as a language itself; they also employ sounds in ways that are not experienced in everyday life. Chanting, drumming, the low vibratory notes of the horns blown during Tibetan rituals, etc. all contribute to nonverbal communication which relies on the active stimulation of the sense organs without necessarily being carriers of meaning.

The point of these forms of ritual is one that challenges the linear patterns of waking thought. From a Tibetan philosophical perspective, for instance, there are two types of "truth": the conventional truth of appearances in which observed phenomena seem to have real existence (are perceived as being real and as existing) and as having essential independent being; and what is called "ultimate truth": a perception that these appearances are illusory, creations of consciousness that disappear during altered mental states when "emptiness" is perceived, and which reappear once the meditative state is abandoned.[13] Thus the ritual space is one in which appearances are treated as playthings or as objects to be manipulated, or even drained of their conventional meanings and usages and assigned different qualities completely. Hence the fantastic-seeming illustrations of the mandalas and other Indian and Tibetan religious art as well as the surreal drawings seen in European alchemical literature. In any coding system letters, words and numbers—digits—no longer have the meanings usually assigned to them; this is even more apparent in analogue systems such as art, music, poetry, and ritual.

In the aggregate, all of these attempts at creating non-verbal states eventually reduce to a single state of nonduality, of what is called *advaita*, and which may be the result of complete harmonic balance between opposite states so that they negate each other (i.e., [+1] + [-1] = 0, or "emptiness," or even "a wheel"). This is the imposition of order upon the chaotic nature of a reality that is composed of appearances which, taken individually, seem to create conflicting emotions of desire and abhorrence, love and hate, beauty and ugliness, etc. Those who are seen as having transcended such states are considered either deeply spiritual by observers, or as robotic or even "alien." A lack of affect often is considered a sign of psychological impairment. Those who test high on the autism spectrum, for instance, but who are high-functioning—those with Asperger's Syndrome—are seen as gifted intellectually in some ways but as lacking the capacity for human contact and response. Isn't that how aliens often are depicted in popular media?

The human being who is credited with having achieved the ultimate state of enlightenment is, of course, the Buddha himself. It was the Buddha—during his first sermon at the Deer Park—who introduced the concept of the wheel as an icon of the religion. In fact, one of the symbols of Buddhism is the image of this first sermon portrayed as a wheel with a deer or many deer in attendance. The footprints of the Buddha also are depicted as having wheels and swastikas: the swastikas symbolize "auspiciousness" in Asian iconography, but also represent the spinning of the Wheel of Dharma, the Wheel of the Law of Buddhism. The wheel is also the primary shape of the mandala, which often is drawn as a circle or a circle within a square (or a square within a circle). This circle is heavily populated with images and colors, usually divided along four quadrants, with various deities, objects, and other ornaments in their appropriate places. In the center of the mandala is often seen a building or a mountain, a literal axis mundi which may represent Mount Kunlun or Mount Sumeru or Mount Kailash: all peaks representing the center of the Earth horizontally and its vertical connection to the polestar.

The fact that mandalas are more often than not divided into four quadrants brings back to mind the cruciform ideogram for "shaman" that we discovered in the previous chapter. In the Indian and Tibetan forms, however, this concept is magnified greatly. The ideal wheel as interpreted by the Buddhist commentators is resplendent in gems, gold and silver. The aspect of the wheel that is emphasized is its ability to shine and reflect light. The center of the wheel—the nave or hub—is described as being made entirely of sapphire, which is the same gemstone that we encounter in the vision of Ezekiel (1:26) as the substance of the throne in the center of the wheeled vehicle. Around the sapphire nave is a silver lining, and then more gems, etc. This is the same wheel that was perceived as flying around the king's tower, and in one Buddhist text[14] we read that it ". . . proceeds through the sky, not very high but just above the summit of trees, so that those who accompany [the wheel] (through the sky) are able to enjoy the fruits, flowers and tender leaves of the trees . . . [it] moves at a height that is neither too high nor too low so that people on earth are able to point out and say, 'That is the king . . .'" This is obviously the description of a vehicle in the shape of a wheel or disk that is observed by people flying through the air and which also boasts passengers who are presumed to be the lords of the earth, the *cakravarti*.

We propose that the mandala, the wheel, and the flying disk are cognate representations for a real object and a real event or events during which this object was seen and was able to enter into the cultural stream. The mandala—in particular the one that represents the Kalacakra initiation—is a focal point for meditation and ritual observance and we propose that this may be due to an effect such a sighting had on an early population.

There may be some modern evidence—albeit circumstantial—that supports this theory. It may provide a clue as to the connection between ancient Asian esoteric systems and the presence of UFOs in a way not imagined by many "ancient alien theorists."

In 1975, famed UFO researcher and astrophysicist Jacques Vallée published his influential book on the UFO influences on humanity, *The Invisible College*. It was widely and warmly received by such notables as US astronaut Edgar Mitchell (a UFO believer) and maverick scientist John Lilly.

On page 197 of that book we see a graph that Vallée created based on the number of UFO sightings in the world from 1947 to 1962. He called it the "Schedule of Reinforcement" and writes:

Figure 3 shows the variations of an external
phenomenon (the UFO manifestations) to which
human society is reacting in various ways. It is
interesting to ask whether this process is not subtly
changing us.[15]

The graph shows various peaks in UFO sightings in
certain years. The years where we see the highest concentra-
tions are 1947, 1950, 1952, 1954, 1956, 1957, and 1962.

Something about those dates triggered a dim recollec-
tion. A look through the available records of the Kalacakra
initiations conducted by the Dalai Lama show two major
initiations involving 100,000 individuals each in 1954 and
1956, in Lhasa, Tibet. These were the first two initiations
conducted by the young Dalai Lama before his forced depar-
ture from Tibet for India after the Chinese invasions, and
they occur in the same two years as important UFO "waves."

This is not to say that the Dalai Lama was in any way
"creating" the UFO sightings or was somehow directly
responsible for them, but such a concentration of people
during an intense ritual experience involving the Kalacakra
(the Wheel of Time) may be connected in some way to the
UFO events in ways we do not yet comprehend. In other
words, as Vallée noted, they may be "subtly changing us."

Not satisfied with these two sightings we looked more
deeply into the record to see if there were any other suggestive linkages between the Kalacakra initiations and UFO
sightings. There were many, but a few stand out as being of
greater interest.

September 1976 was the month of the famous Tehran, Iran, UFO sighting in which two Iranian jets were involved. The incident was the subject of a US Defense Intelligence Agency report, and the Iranian Ministry of Defense openly stated the cause was extraterrestrial. That same month, the Dalai Lama conducted a Kalacakra initiation in Ladakh, India, for 40,000 people.

December 1996 was the sighting of a UFO by video taken aboard the STS-80 Space Shuttle Columbia. That same month, the Dalai Lama conducted a Kalacakra initiation before a huge crowd of 200,000 people in West Bengal, India.

The sighting of December 1985, however, is *very* suggestive. That month the Kalacakra initiation was conducted before another large crowd of 200,000, this time in Bihar, India. That same month saw the famous *Communion* experience of alien abductee Whitley Strieber which became the subject of the book and the movie of the same name.

This episode is so well-known among UFO enthusiasts and the general public alike that it needs no detailed description here beyond a brief summary. Strieber, a Hollywood screenwriter, was at his cabin in upstate New York when he had an experience that can only be described as a kind of abduction from his home in the middle of the night by beings that were not human. The substance of this event was recounted in his bestselling book, *Communion*, and eventually became a film starring Christopher Walken as Strieber. Strieber has since suggested that the experience had spiritual dimensions, as well as perhaps a psychological basis extending back to the time he spent as a child in San Antonio, Texas, near the Randolph Air Force Base where the Nazi scientists involved in aviation medicine—brought over under the infamous Operation Paperclip—were based. This experience changed Strieber's life forever and he has since been a tireless advocate of the UFO phenomenon as worthy of serious attention and as evidence of a spiritual change taking place on the planet.

Vallée agrees that the UFO experience is changing us in "subtle ways." He links the experience to a kind of control operation by a force or forces unknown. In fact, in *The Invisible College* he shows comparisons between UFO experiences and religious experiences,[16] thereby drawing some similar conclusions.

That religion often functions as a control mechanism is not to be doubted; but what if the UFO experience stems from the same source, as we have been insisting all along? To go even further, what if our understanding of the mechanisms of both religion and UFOs is so flawed we cannot entertain the possibility that we can take charge of the mechanisms ourselves?

Those who do, or who profess that they do, are often isolated from society and ostracized. Certainly, actions that are not consistent with acceptable "human" behavior will result in exile or self-exile from the "haunts of men" as one dreams "the dream of Heaven" to quote Lord Byron. This occurs either naturally or accidentally—as in the case of the Siberian shamans or the psychologically impaired—or deliberately, as in the case of those who undergo strenuous initiations into spiritual disciplines that require abandoning one's natural inclinations in favor of direct experience of the Phenomenon. The proliferation of UFO cults—especially in the last hundred years or so—is evidence of the degree to which many individuals understand the religio-spiritual implications of sightings and experiences. While the UFO cults often are based on accidental sightings and experiences, the more formal initiatory groups are predicated on the belief that these sightings and experiences can be deliberately courted, if not caused. This is an aspect of the Phenomenon that largely has gone unnoticed because of the assumption that UFO contact and spiritual contact are fundamentally different.

In fact, one of the core documents of the Kalacakra Tantra tradition is the *Vimalaprabhā*: a commentary on the Kalacakra Tantra that is supremely influential. A translation of one section—entitled "A Summary of the Vajrayoga"—was published in 2000 and contains a revealing statement. After a brief discussion of the two types of knowledge—conventional and ultimate—that we discussed above, it continues:

> It is directly perceived: an appearance of his own mind manifests in the sky for the yogin, just as a divination appears in a mirror for a maiden.[17]

And again:

> . . . it is not imagined by one's mind—it is directly seen, an object of experience. It is all aspects; it

originates from the sky. It is completely good,
the total cognitive faculty. . . . It has completely
abandoned logical reasons and examples.[18]

The yogin, of course, is the practitioner of yoga: in
this case, *Vajrayoga* (the yoga of Tibetan Buddhism). After
practicing yoga and meditation within the context of the
Kalacakra initiation, the yogin then sees "an appearance
of his own mind in the sky." If we multiply this by the
hundreds of thousands of people taking the Kalacakra initiation at a single time, as noted above, then the possibilities for manifesting an appearance "in the sky" is increased
exponentially. The mass Kalacakra initiations—particularly
those outside of Tibet and outside of Asia generally—began
in the same era as the modern UFO sightings. While much
more research and analysis is needed before we can propose
a direct cause-and-effect relationship (if that is ever possible) the synchronicities mentioned above should at least
give one pause. The Kalacakra initiations would not have
been the only UFO precursor, of course, even allowing for
a connection between the Phenomenon and an esoteric
or religious practice. Other operations intended to alter
consciousness and achieve states of awareness that extend
normal human experience may also effect sightings and, of
course, there is a rich history of these sightings going back
thousands of years: long before the Kalacakra Tantra was
transmitted circa 700 CE.

The statement that it "completely abandoned logical reasons and examples" is as good a description of the

Phenomenon and its presence throughout recorded history as any. The Phenomenon defies logical explanation. It appears and disappears, makes impossible aerial maneuvers, ignores gravity, interferes with electronics (or not), and is blamed for everything from alien abductions to cattle mutilations and California cultists. Sometimes the sightings are of disks, at others of triangles, at other times they are cigar-shaped or—as in the case of the Great Airship Sightings of 1897—they are steampunk machines with crazy gears, weird spotlights, and sails. That there is a history of strange things seen in the sky is not to be doubted; that this Phenomenon has persisted for so long, under so many different circumstances and in so many varied forms, shapes and colors, demands some kind of serious attention and analysis, which so far generally has been lacking. The contours of this Phenomenon may lie in the various ways religions and cultures have been influenced by it and have adapted it towards applications that are unbelievably idealistic, romantic and ambitious: immortality, transformation, celestial flight. The rituals and meditations of the Tantric yogin have as their goal the same as the mummification rites of the Egyptians. All of these represent efforts to transcend our natural condition, and the UFO represents the ultimate in that transcendence. So it begs the question: which came first, the UFO or religion? Or more specifically, esotericism?

And why is matter (and especially the flesh) considered evil by so many of the world's religions, a prison for the soul? What does it tell us about the celestial origins of humanity?

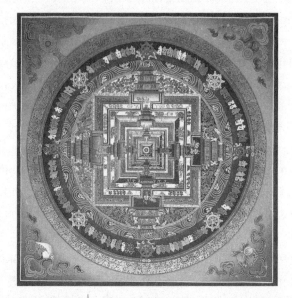

This is a depiction of the mysterious land of Shambhala, an integral part of the Kalacakra Tantra: the core scripture of Tibetan Buddhism. This particular image comes from the Sera Monastery. Like the magic circles shown in previous chapters, this is a series of concentric circles around an interior that is divided into four quadrants. There is writing around the circles and eight-rayed wheels or *cakras* in each of the four quadrants. The center image is of a stepped pyramid with a temple or shrine at the top. Thus the structure is consistent with those found in Sumer, Mexico, and elsewhere.

The Kalacakra mass-initiations conducted by the Dalai Lama from 1947 to 1962 corresponded with spikes in UFO sightings, as mentioned in this chapter.

PLANET EARTH: THE HOT LZ

Was there a period of temporary insanity, through which the human mind had to pass, and was it a madness identically the same in the south of India and in the north of Iceland?

– Max Muller, *Comparative Mythology*[1]

I had always been puzzled by man's use of the word "God." When the Bible says that "God spoke to" or "He spoke to God," I had always assumed it was a figure of speech. . . . But if humans were reached by beings from space who were so superior that man in his ignorance identified them with gods (the Elohim of the Bible), or God, then the literary allusions become clear.[2]

– Andrija Puharich, *Uri*

PERHAPS THE CULTURE THAT HAD THE SINGLE GREATEST influence on the development of Western civilization is the ancient Greek. We derive our ideas about politics and democracy from the Greeks. The Greek alphabet is used in our mathematics and sciences. We name celestial objects after the heroes of Greek mythology. The writings of Plato and Aristotle have exerted tremendous influence over philosophy. Euclid in geometry. Aristophanes, Aeschylus, Euripides in drama. The list is endless. Even Freud employed Greek ideas and motifs in developing his ideas about human psychology, such as the Oedipus Complex.

The translation of the Bible known as the Septuagint is written in Greek. Many of the Gnostic texts are written in

Greek. The very word *gnosis* is itself a Greek word. Greek ideas about the soul, the spirit, knowledge, intelligence, etc., permeate theological speculations in the West both in terms of traditional religion and those of mysticism. In fact, the word "mystic" comes from the Greek and also means "secret."

Yet, it was the father of comparative mythology and one of the first Western experts in Indian religion and Asian studies, Max Muller, who acknowledged that there was little difference between Greek mythology and the beliefs of the "heathen world."[3] Thus, absent any "heathen" objections, it is profitable to take a look at the Greek ideas about the creation of the world, of humans, and of the relations of humans to the gods, in order to come to some interesting and hopefully clarifying conclusions. Indeed, some of the strange hieroglyphic characters that have been reported as being engraved on UFO artifacts bear strong resemblance to Greek letters, which is suggestive of some sort of psychic connection or resonance if nothing else.[4]

Muller asks if the mythologies of the world are evidence of a time of temporary insanity, one which affected the entire globe. As an alternative, he proposes that myths represent actual events and should be taken, if not literally, then as close approximations of historical truth. It is but one step from there to the realization that peoples around the world were talking about a celestial "intervention" on the Earth at some point in prehistory. In fact, this would be the only scientific explanation for this uniformity of belief and one that is consistent with the data contained within the myths themselves.

The Birth of the Giants

> Oedipus, Narcissus, Orestes, Kronos devouring his children, Prometheus the fire-thief, are the psychically richest yet most economic crystallizations of elemental impulses and configurations in the unconscious and subconscious fabric of the race and of the individual. It is in these "prime" myths that our consciousness finds its ever-renewed homecoming to the opaque comforts and terrors of its origins . . . the uncanny and the daemonic.[5]

About 700 BCE, the Greek poet Hesiod composed his *Theogony*. This was an account of the birth of the gods as well as the creation of the Earth and of human beings. As with many of the Near Eastern accounts, according to Hesiod the original state of the cosmos was Chaos. Out of this Chaos was formed the Earth, and the Earth and Chaos mated and began to produce offspring.

Tartarus was born: the Underworld. Then Eros, the god of desire. And then a whole generation of dozens of fantastic creatures, some of whom are familiar as popular images in books and film such as the Cyclops, and the Titans. Kronos (Saturn), one of the Titans, was born and became something of a problem child.

Kronos actually takes command of the entire Cosmos, and in order to ensure his position and remove any possibility he will be deposed he devours his children by his consort Rhea. He castrated his own father, Uranus, whose blood then fell to the Earth and engendered the Furies, the Meliae and the Giants. The testicles of Uranus fell into the sea and engendered Aphrodite, the Goddess of Love. (Go figure.)

That the blood of Uranus fell to the Earth and created the Giants seems to echo the biblical theme of the Nephilim. The Meliae actually refer to the race of human beings who appeared during the Bronze Age. They had been preceded by a Golden Age race and a Silver Age race, and would be succeeded by the Iron Age (our current Age according to this account). The Bronze Age Melians and the Giants were warlike and eventually destroyed by Zeus who caused a Flood to erase them from the face of the Earth in an echo of the biblical account. The current human race was born of Deucalion and Pyrrha, who were warned by Zeus of the coming Flood and who found themselves beached at the top of Mount Parnassus when the flood waters resided. Obviously, the parallels with the biblical accounts and even with other Near Eastern accounts is so strong as to suggest that they stem from the same source.

What is often not remarked, however, is the similarity between the accounts of the human race having been created from the blood of the gods which also contributed to the creation of a race of Giants (*Gigantes*). The fact that the Greek accounts mention the Gigantes as having been

created in the era preceding the Flood—such a clear parallel to the biblical account in Genesis—seems to justify the
interpretation of Nephilim as "giants." There is controversy
over this term, however, with some scholars claiming that
the association of the Greek word *gigas* with our understanding of "giant"—i.e., a being of enormous size—came
later. It was believed that *gigas* meant "one born of earth"
as the Giants were the offspring of Gaia, the Earth. (Hence
Gaia: *gigas*.) It is now believed that the Greek word *gigas*
has its origins in an unknown non-Greek or pre-Greek language and context. In other words, the actual meaning of
gigantes and *gigas* is as obscure as the actual meaning of the
word *nephilim*, two words in different languages that mean
the same thing! This may be further evidence that whatever was represented by these terms was a real being whose
actual name was interpreted or translated by the Jews and
the Greeks based on their own contact with them, or based
on a common source at some point in prehistory.

In any case, the descent of humans from the blood of
the gods—a blood that also produced monstrous abnormalities—is a constant theme from ancient Israel to ancient
Greece to Sumer. There is, in addition, another constant
theme: that of a conflict between celestial forces that had
ramifications here on Earth. This conflict has a name in
Greek: *Gigantomachia*.

It should be pointed out that the Greeks did not cherish
a close relationship with their gods. In fact, they preferred
to be ignored by them. The gods were not predictable; they
acted with seeming irrationality amongst themselves and

in their interactions with humans the humans invariably suffered. The Greeks felt that humans were the playthings of the gods, which might be a way of saying that humans are vulnerable to the actions of fate or nature or similarly uncontrollable forces. This sense that the gods and humans had a difficult relationship is mirrored in many Western traditions. Certainly the Books of Genesis and Exodus (as much of the Bible) is replete with cases of God punishing humans, both Jew and Gentile alike.

The Giants of Greek mythology were warriors; they were not necessarily tall, as the illustrations of them found in Greek pottery and other sources indicate that they were more or less of human proportions, but they were enormously strong and imposing and were adorned with armor made of bronze and wielded heavy weapons. They cared only for battle and nothing else, and presumably it was their reign of terror on Earth as well as their antagonism towards the gods of Mount Olympus that encouraged Zeus to have them destroyed.

Before that happened, however, the Giants engaged in battle with the other gods in a conflict that raged for long periods of time. According to the Latin poet Ovid in his *Metamorphoses* (1:151-162), the Giants angered the gods of Olympus due to their piling mountains on top of mountains in an effort to reach the stars: an obvious parallel to the biblical story of the Tower of Babel. In another version, the Giants were created by Gaia because her previous brood of offspring—the Titans—had been defeated by the gods in an earlier divine conflict and she desired to avenge them by sending the Giants against Olympus. Prometheus, who will figure later in this story, was one of the Titans who was punished by Olympus and who was also responsible for creating the human race.

The war between the Giants and the gods of Olympus for control of the universe was known as the *Gigantomachy*: the "war of the giants." It was known and discussed as early as the sixth century BCE and was the subject of much speculation. There is a tendency among scholars to equate myths with natural phenomena, saying they are a way for primitive peoples to explain such things as volcanoes, earthquakes, etc., by anthropomorphizing them to some extent. The same was true of the Giants who were associated with volcanoes. According to Hesiod's *Theogony*, virtually every natural phenomenon or object was named after a god or goddess: Helios for the Sun, Selene for the Moon, Eos for the Dawn, etc. The interactions between the gods, however, were not as easily associated with observable phenomena. For instance, how would one interpret the birth of Aphrodite, the Goddess of Love, from the severed testicles of Uranus, as a natural phenomenon? It has the hallmarks of a story that is deliberately encoded; and without knowing the context it defies interpretation.

The Gigantomachy is one such account. To insist, as do some scholars, that it was invented in order to "explain" the origin of volcanoes (for instance) strains credulity. It is far simpler to recognize that the legend is an account of a real event. It was such a forceful event, in fact, that it was memorialized for centuries and formed the substance of a variety of representations from the Greeks to the Romans and beyond. It has been interpreted as the triumph of order over chaos, as the Giants lost their war against the Olympian gods and Zeus then cleansed the Earth of the Giants in a Flood and rebooted Creation. It has also been suggested that it refers to the dominance of the invading Greek people over a barbarian society, but the account runs so closely to the biblical version that it must reflect more than an isolated Greek phenomenon.

The Giants were "born of the Earth" and from the spilled blood of the gods. They represented Matter in its most basic form—earth and flesh—that had been animated by the divine essence, by Spirit. They revolted against the Olympians, which can be construed as the revolt of Matter against Spirit, or the revolt of a prehistoric race of human beings against the gods. It is the contention of this project that both the literal revolt and its philosophical interpretation are represented in the legends and myths; that concrete experiences are not perceived as phenomena extraneous to consciousness but that there is always a psychological component. Therefore, in this case, there was an experience of monstrous or abnormally developed creatures—human or quasi-human—in conflict with a race of superior beings and who were defeated by them, and this event or series of events preceded what we know as the Great Flood. That experience was internalized to an extent mediated by culture and language so that an interpretation of it took place almost immediately: a narrative, a creative linking of events, personalities and ideas such as we find in histories, biographies, etc. The defeat of the monsters or giants was justified on the grounds that they represented chaotic, chthonic forces that resisted all attempts at orderly development, development that was represented by the gods of Olympus and specifically by Zeus, the father of the new race of gods.

Where were the humans when this conflict was taking place?

Strangely, the epics are silent on this obvious discrepancy. How would the Iron Age humans know what had transpired to their Bronze Age ancestors? We are told by the Greeks—as the Jews were told before them, and as the Sumerians were told before them—that there were some humans who escaped the Flood that destroyed the Giants and it's from these humans that the present race is descended. We saw this same story repeated in the Vedic legend of Manu and Matsya in the previous chapter. There is thus a kind of memory of these events that has been preserved and it is possible—highly probable, in fact—that the present human race began as an abnormality itself. As a kind of monster.

The Flood destroyed most of humanity, according to the legends. That would indicate that most of humanity belonged to the race of Giants or Nephilim, that they were the norm and not the exception. The fragments of humanity that were saved—either by Noah or by Deucalion, etc.—were a minority, a handful, an exceptional gene pool that was not giant or Nephilim. In other words, *we* are the mutation.

In fact, we know this, although perhaps not in so many words. We treat the planet we live on as an alien place, an external object that is not intrinsic to our own nature. We objectify it in so many ways. This is not meant as a value-judgment or a criticism, but merely as an observation. We stand in relation to the planet the way a chemist stands in relation to a laboratory: a place of test tubes, flasks, Bunsen burners, chemicals, tubing, measuring devices both analogue and digital, notebooks filled with equations. The chemist is the only moving component in a room filled with stationary objects. The planet, like the laboratory, is something *out there*, something held at arm's reach. It is not part of us. We are just passing through.

It is consciousness that creates this division between humans and the planet, and it is precisely consciousness that cannot be measured or weighed or otherwise understood. To a school of philosophers that came out of the Greek cultural milieu, consciousness is trapped within the human body and is not happy about it. This school is known as Gnosticism.

▼   ▼   ▼

In 1945, towards the end of World War II, a cache of ancient documents was discovered in Egypt. Known as the Nag Hammadi Scrolls, these represent some of the earliest examples of Gnostic writings. The Gnostics were a mystical sect that combined elements of Judaism, Christianity, and other religious and spiritual traditions—including Egyptian—into a single belief system that characterized the world as a battleground between the forces of Light and Darkness. This spiritual conflict is taking place within every human being, as humans are made of both Matter and Spirit. It was the goal of Gnostic practice to separate the spiritual part of humanity and have it return to the heavens.

Their interpretation of biblical literature was consistent with this belief, as we saw at the beginning of this book. To the Gnostics, the Serpent in the Garden of Eden was God and the Being that told Adam and Eve not to eat of the fruit of the Tree of Knowledge was the Demi-Urge: a being who had usurped God's role in Creation. The Gnostics valued spiritual knowledge above all other kinds, hence the name by which they were known: the Greek word *gnosis* which means "knowledge." To the Gnostics, God—the true, transcendent deity of Light—was an Alien God, the cosmos was the realm of Darkness, and humanity was in chains:

> The deity is absolutely transmundane, in nature alien to that of the universe, which it neither created nor governs and to which it is the complete antithesis: to the divine realm of light, self-contained and remote, the cosmos is opposed as the realm of darkness. . . . The universe . . . is like a vast prison whose innermost dungeon is the earth, the scene of man's life.[6]

This is the philosophical basis for Manichaeistic duality which, as we have seen, used the *Book of Enoch* and the *Book of Giants* as inspiration: books that were eventually discovered among the Dead Sea Scrolls in 1947. Taken together, the Dead Sea Scrolls and the Nag Hammadi texts shine a light on Gnostic and early Christian and Second Temple Jewish beliefs and doctrines, but they also represent ideas that were current in the Near East centuries earlier. The *Book of Giants* even mentions Gilgamesh, the hero of the Sumerian and Akkadian epic poems, and reprises the biblical story of the "sons of god and the daughters of men" but with more detail, including allusions to the giants having bad dreams and asking Enoch to interpret them.[7] This document—heavily fragmented—also mentions that the angels (the "sons of God") were engaging in sexual intercourse with human women and also with animals, and that in the process they begat "giants and monsters."[8] (It is interesting to speculate that this may be an attempt to explain the paintings, statues and bas-reliefs of Egypt and Mesopotamia depicting half-human, half-animal creatures, such as the animal-headed gods of Egypt or the *apkallu* of Mesopotamia.)

This example of widespread genetic abnormality was disastrous enough that the memory of it lingered for hundreds if not thousands of years. It may have contributed to the horror the Gnostics felt for works of the flesh and the reproduction of matter, perhaps believing that the possibility always existed that monsters would be produced, that humanity would degenerate and that any possibility for advancement of the race could be dashed again and the Earth plummeted back into another Flood, or worse.

Whatever the reason, the Gnostic worldview was sharply divided between the created world which was considered the world of darkness and evil, and the world beyond or outside creation: a world of light where the darkness could not penetrate. Human beings were imprisoned in *this* world, which provides a Gnostic context for some of the Sumerian speculations of Zecharia Sitchin and other ancient astronaut theorists who put forward the idea that humans were created as a slave race to mine gold, etc. What is compelling about the Gnostic version is the idea that all of creation is evil: the gods, the angels, the devils, the planets, the stars, etc. It is *all* evil, and is to be transcended. The real God of the Gnostics is a god of Light and is completely removed from creation and, in fact, had nothing to do with it in the first place. This is, in a sense, true religion for it is a religion that goes far beyond the ancient alien theories, all denominational concerns, all theological hair-splitting, and all the themes we have been exploring, to arrive at a conclusion that is deeper and more profound than the Cargo Cult approach we have been taking so far. It is closer to the Buddhist conception that all is *maya*, illusion, and that even the gods are to be distrusted as mere images summoned up by consciousness. The Buddhist seeks the Void in order to find enlightenment; the Gnostic seeks to escape the world of matter and thus of all images and appearances in order to reach the Light. To the Buddhist, all appearances and images are equally valid and equally unreal. The Gnostic has a similar point of view, and would agree with the Buddhist that "all is suffering." They may put forward different reasons why this should be so, but they are in agreement that humanity is in chains, a slave that yearns towards freedom and liberation.

We feel that this concept is both literal *and* figurative. We feel that the stories of beings foreign to this planet using its chemical elements and something of its own genetic material to create human beings has some validity in the real world; it just begs a coherent explanation based on evidence. This eventually may be provided, either through research into directed panspermia or in ways closer to the so-called "mythic" accounts. People all over the planet feel this, even if they do not give voice to the feeling the way we are doing here. What evolutionary purpose is served by this nagging sense—revealed in countless spiritual, religious and historical accounts over the millennia and from all over the world, not to mention in the world's literature, music, and art—that we don't really belong here, that something is wrong with us, that death is an absurdity, and that our home is in the stars? Why did our evolutionary line survive the Great Flood, if it was only to be removed even further from our point of origin, from the knowledge of our relationship to the gods?

Or is this feeling, as Max Muller described it, nothing more than "temporary insanity"? Is this planet not really a prison, but an asylum?

Alien Nation

> Alienated? Of course I'm alienated. I'm an alien!
> — Alan Arkin as Prof. Simon Mendellsohn
> in *Simon* (1980)

One of the insurmountable problems facing any society when alien contact occurs will not be the psychological effect of this contact on the human population but the urgent need to understand the psyche of the aliens themselves. From the point of view of psychology, there is no conceivable way one would be able to assess the mental health or illness of an alien species. The planet could be invaded by the alien equivalent of psychopaths and there would be no way for us to know.

The giants and monstrosities recorded in the *Book of Giants* as well as the slim account of the Nephilim in Genesis need not be understood solely in terms of physical abnormalities. Modern psychiatry understands that physical defects may have corresponding mental effects, or effects that are the result of the poor social interactions caused by physical defects such as being shunned by other human beings due to perceived ugliness or some disturbing physical characteristics, etc. The monstrosities of the pre-Flood era may have been psychological monstrosities as well as, or in place of, physical abnormalities. Certainly, from a modern perspective, the actions of the "giants" could be interpreted as sociopathic. The sexual proclivities of the angels—their parents—would be construed as deviant and criminal behavior today: incest, bestiality, etc. Considering that the angels—the *bene elohim* of Genesis—are the "sons of God," it stands to reason that there is a genetic circumstance which contributes to this psychopathic behavior. We assume that the *bene elohim* were directly created by God, according to the biblical version; their mating with humans, a kind of celestial miscegenation, caused the birth of giants and monsters. The humans on their own did not seem to produce hideous offspring. It was only when they mated with angels that the dysmorphism occurred and it seems to have survived in isolated pockets of the gene pool after the Flood. Giants such as Goliath still existed, and there are numerous other references to giants in the Bible in the post-Diluvian period.

Was the biblical gigantism the result of a chromosomal factor, such as may be caused by consanguinity? In other words, was the blood of the "sons of God" so similar to the blood of the "daughters of men" that a genetic predisposition to gigantism was exacerbated? In other words, was it a case of inbreeding?

Consider the Sumerian and Babylonian creation epics in which human beings were fashioned from the blood of the slain Kingu. If we take this story literally, or as far as a literal interpretation can take us considering the lack of details, then human beings are genetically related to an ancient race of monstrous beings. A certain degree of genetic diversity among humans is assumed to have prevailed for quite some, in accord with the way geneticists trace all human origins back to an original "Eve" somewhere in Africa who was the "mother of us all." However, at some later date, descendants of Kingu or at least of the race of Kingu and Tiamat returned to Earth and mated with people who were their own relatives by blood. These were necessarily consanguineous relationships, i.e., the participants shared the same blood.

Regardless of how humans were created—through a deliberate act by some divine, celestial or alien being or through the accident of panspermia or even by directed panspermia—there was enough genetic similarity between the humans and the "sons of God" that (a) sexual intercourse was possible between them in some fashion and (b) offspring were produced from these unions. Too much genetic similarity in human beings who procreate among themselves contributes to birth defects. It is rare for gigantism to result from inbreeding, however, although in rare cases it is possible; the more likely result would be dwarfism or other genetic abnormalities. Along with the physical symptoms some mental or emotional symptoms might also present themselves, such as schizophrenia (which has a genetic component), paranoia, aggression, etc.

What we are dealing with here, however, is not typical human inbreeding. If there was an ur-chromosome, for instance, some Rosetta Stone of the human genome that can be traced through all human beings and is present in animals as well, it might help identify an extraterrestrial or non-terrestrial source, particularly if a mutation can be identified at some point across some or all species that would signify an external tinkering with the process. We will look at this possibility in more detail in Book Three. For now, however, let us analyze what ancient texts are telling us about the psycho-biological or psycho-spiritual makeup of the gods, angels, demons, and other nonhuman beings in order to understand something of our own human condition.

In order to do this, we have to look across religious denominations. Using only the Bible as a source would be counter-productive. If there was alien contact—something identifiable as extraterrestrial or non-terrestrial—then we have to assume that it was experienced by all cultures whose texts, scriptures, or oral traditions contain references to celestial beings, powerful and non-human agents with supernatural powers, etc. This was either an observable event, such as a UFO event or accompanying a UFO event and seen by multiple witnesses, or it was something subtler such as an alien abduction experience or what J. Allen Hynek called a "close encounter of the third kind." The reality is probably somewhere between these two, for the following reason:

Spiritual sources often refer to celestial beings and celestial events. If the paleocontact, whatever it was, was purely personal and subjective like a vision or hearing voices, etc., then the observable celestial component would not be present or associated in any way. We would have a "God spoke to me" experience and nothing more. The large number of sources that describe beings coming down from the heavens or originating in the heavens or ascending to the heavens, however, means that there is a definite

UFO component present in the texts we have been discussing. There is another possibility, of course, and that is that the UFO component occurred long ago and that the experiences people have been reporting in the millennia since then have been of a series of remote contacts conducted over a communications medium we do not understand. In other words, the initial contact was like employees meeting the boss for the first time; the succeeding contacts are the aliens "working from home" or maybe video conferencing.

This would be so whether our alien component is purely genetic, and thus occurred long before we were conscious (and thus our memories of the alien aspect of our evolution somehow are genetically based) and before the evolution of *Homo sapiens*, or there was contact at some point during our conscious development. In fact, it is likely a combination of the two scenarios; i.e., our genetic inheritance is of stellar origin *and* at some point along our evolutionary timeline we were visited by the forces that were responsible for the "seeding" of our planet.[9] That would explain the fact that recorded accounts of these various contacts throughout the world involve beings with whom we have at least some characteristics in common, but with whom we have many differences as well. It is similar to white Europeans visiting China for the first time in modern history: there were many similarities biologically between the two races but no common language, no common culture, no common history. Indeed, as far as the Chinese were concerned, Europeans were not humans at all but devils. We contend that contact between humans and the alien or extraterrestrial or non-terrestrial beings would have had (and still do have) much of the same flavor, only more so. The aliens are alien to everyone on the planet, equally. They are the world's "foreign devils" and it is possible that this facile characterization is not very wide off the mark.

Anthropologists have noted that many peoples refer to themselves as "human" while reserving other terms for those who are not of the same clan or tribe. In some cases, the word for "human" and the word for a member of the tribe is the same word, implying that those who are not tribal members are not even human beings. This is taken to its logical conclusion in the Chinese (and Japanese) scenario in which ancient perceptions of foreigners were equated with non-human, quasi-supernatural beings, like devils or ghosts, and these terms became enshrined in the language with such phrases as "white ghost" and "foreign devil." What precedence was there for this type of reaction to beings who were clearly similar in virtually every respect to native persons—two arms, two legs, a head, a torso, etc.—with certain exceptions such as color of skin or hair, etc.? There is no evolutionary benefit to be had from inbreeding; in fact, just the opposite. This should make the members of other races more attractive for purposes of reproduction, but paradoxically they were not seen as members of the human race but as animals or ghosts or devils, and for that reason unfit for reproduction.

So the question remains: at what point did the human race learn to be suspicious of other beings who appeared human but who were, in fact, something other? And is this social taboo against breeding outside of one's clan, ethnicity or race a device to ensure genetic weakness in the human race and its vulnerability to an outside force?

▼   ▼   ▼

Alien abductee reports often include scenes in which the human victim or abductee is experimented upon sexually. There is an abnormal preoccupation with sex organs, seminal fluid, and impregnation. Based on the previous chapters we can see immediately that the ancient accounts of celestial beings mating with human beings is a standard motif in many of the world's religions. What is taking place in the present day is a reprise of the ancient legends and demonstrates that the Phenomenon has remained steady throughout the millennia. There is talk in the UFO community today about "hybrids": humans born on this planet of both alien and human DNA. This is a modern interpretation of the "sons of God" and "daughters of men" scenario of the Bible, and of the Titans, the Giants, and the Melians of ancient Greek mythology, as well as the origin of human beings as recounted in the Sumerian and Babylonian epics. If we posit an alien abduction taking place in ancient Israel, for instance, it would have been recorded much the same way that Genesis recorded angels mating with humans and creating monsters. It is the same Phenomenon, seen through different eyes in different times and places on the planet and with minor modifications according to culture and language.

The prevalence of antinomian sexual practices taking place among those religions that were created in the past two hundred years or so is further evidence of the relationship between alien contact and sexuality. The plural wives of the early Church of Jesus Christ of Latter-Day Saints is but one example; Joseph Smith, Jr., had made contact with a non-terrestrial being through a series of magical operations, out of which the Book of Mormon was produced. Scientology, considered by many to be a "UFO religion" with its science-fiction mythology and its occult origins, also has had its share of exotic sexual relationships, especially with regard to its founder, L. Ron Hubbard, and his several wives, including one case of bigamy.

In this connection it also should be pointed out that we humans are inordinately fond of disaster scenarios. We are constantly making films that depict in graphic detail the destruction of our cities, our countries, and the entire planet. Disaster films—whether of natural disasters such as tornadoes, hurricanes, tsunamis or earthquakes or of terrorist plots or nuclear accidents, etc.—are a troubling staple of Western movie-making. They are being produced in the same post–World War II era as that other popular speculative genre: films with UFO or alien themes. In some cases, they are the same film (such as *Independence Day*, or *The Day the Earth Stood Still*). We are producing images that replay Genesis over and over again: one film about aliens and UFOs, followed by one about the Earth being destroyed. This is a kind of modern scripture: a myth endlessly repeated, re-enacted as ritual drama, so is there any reason to be surprised that there are modern UFO religions being created every decade?[10]

However, even as these twin fears of alien invasions and natural disasters trouble our dreams there is the corresponding development of modern technology and especially of modern weaponry. The atomic bomb detonations of 1945 were traumatic globally. They signaled the possibility of destruction on a "biblical" scale, to such an extent that ancient astronaut theorists began to interpret the destruction of Sodom and Gomorrah in terms of nuclear weaponry. While some may say we have stagnated in terms of spiritual progress (by which metric? how do we measure such an intangible?) there has been undeniable progress on the technological side. We are now capable of wiping ourselves off the face of the Earth and in the process making it uninhabitable for centuries. This is a power about which the old time prophets could only dream, an ability that used to be reserved for the imagined power of God alone. We no longer need to wait for God to visit a deluge upon the Earth. We can do that ourselves. As well as pandemics, famines, pillars of fire, swarms of locusts. We have chemical and biological weapons that would make the plagues of Egypt in Exodus look like the common cold.

What prompted us to develop this capability?

The Indian epics—the ones usually cited as evidence of the "flying machines" of ancient legend—such as the *Ramayana* and the *Mahabharata*—discuss celestial conflict in a matter-of-fact way. Many Indologists believe that the *Mahabharata* is an historical record: that the events described in it are accounts of real conflicts that took place thousands of years ago involving human families or clans fighting over territory on the Indian subcontinent, perhaps the much-debated "Aryan Invasion." Yet, the gods are always present in some form such as in the *Bhagavad Gita*, the most famous section of the *Mahabharata*, in which Arjuna and the god Krishna have a spiritual discussion on the eve of battle. So it is legitimate to ask: to what extent do the gods involve themselves in human affairs, directly or indirectly? In the case of the Bible as in the *Mahabharata*, the *Iliad* and the *Enuma Elish*, to a very great extent indeed. There is virtually no political or military event of any importance in which the gods are not involved. There seems to be no human agency, no human autonomy, in any arena of experience except perhaps where making the appropriate sacrifices to the appropriate gods on the appropriate days is concerned. Humans regard the gods as vast, capricious forces that can be manipulated to a certain extent but never controlled.

This idea that the gods were involved in all important civil (as well as spiritual) aspects of human life began to lose its allure about the time of the scientific revolution, at least in Europe and to some extent in North America. That did not stop the creation of new religions, however, with prophets who had spoken directly to the gods (such as the aforementioned Joseph Smith, Jr., as late as the nineteenth century). Gradually, human beings began to take responsibility for their own affairs—from wars and revolutions to economic progress and colonization—and the role of

divine and demonic forces became relegated to purely spiritual matters, like a crazy old uncle removed to the base-
ment of the ancestral home. Churches as institutions were
still quite powerful, but any reference to the role of God
in human affairs was seen as purely *pro forma*, a polite nod
to the crazy uncle rather than putting him in charge of the
books. Religion became a matter for personal belief rather
than public acknowledgment, except in those cases where
mentioning God would guarantee a few more votes.

Eventually, though, God would become irritated with
this cavalier treatment and would decide to remind the
civil authorities—governments, armies, physicists, spies—
that there are more things in Heaven and Earth than are
dreamt of in the pages of the Congressional Record. God,
gods, angels, demons, jinn, spooks, sprites, elves, all that
October Country race of paranormal beings that always
had been associated with the generation of humans and
the creation of the cosmos itself—a race pushed aside, or
greeted with derision and ridicule—would erupt, invade,
explode on the scene in ways that not only the gullible and
the credulous would entertain, but which would tax the
imaginations and the resources of governments and armies
once again, just the way they had during the Egyptian
plagues and the vaporizing of Sodom, the devastation of
Lanka and its demon-king Ravenna, and Plato's fabled,
doomed, sunken Atlantis.

They would fly over the world's capitals. They would
zigzag over nuclear plants and missile silos. They would
wander, red-eyed and gaunt, through the woods of West

Virginia. They would swoop down on screenwriters in upstate New York.

And they would joy-ride through the skies over the southwestern United States like sailors on shore-leave or marines on a three-day pass, their rented T-Birds careening madly through saucer stop-signs and interstellar intersections, whistling at the girls, knocking down mailboxes and howling at the Moon, drunk with oxygen and possibilities.

And once in awhile, like a Chinook or a Huey slamming into a Vietnamese rice paddy, or a hot-rod doing backflips over the median on Highway One, they would *crash*. And when they did, when those little alien bodies flew out the windshield, their sekret machine ripped apart by gravity and hard-baked sand, it suddenly was not so easy to ignore the Hand of God in the world. God, alien, ghost, vampire, whatever supernatural agency one prefers, it was suddenly *there*: part of the real world, intruding into the *weltanschuung* of the armed forces, of newspaper editors and the six o'clock news.

They have it wrong, the ancient astronaut theorists and the experiencers and the abductees. Ezekiel's Chariot was not a flying saucer. That's projecting our current experience backwards into time, stretching to make the details fit like the elastic waistband on a cheap suit. Flying saucers, triangular craft, cigar-shaped craft . . . lights in the sky . . . they *are* Ezekiel's Chariot projected forward into the present. Every generation will interpret biblical phenomena according to their own lights, their own technologies, desperate to make science and religion match up in a nice, neat existential package. We all want to say "See! The Bible is based on truth!" But it won't work. It never does. What is happening is not confined to either side of the great binary system of science-religion, or fact-fantasy, or normal-paranormal. The paranormal is infused in the normal; it's inextricable. It's the engine of creation, of reproduction, of sex and love, and the light at the end of our tunnel vision. It's saying "See! Science is based on truth!" To understand the Phenomenon is to grasp the *advaita*, the non-duality, of modern experience: a non-duality that encompasses more than being and non-being, of god and not-god, but of war and peace, of mind and matter, of flesh and spirit, of aliens and men. This is the temporary insanity of which Max Muller wrote so long ago.

One of the facts about the conflicts between the gods that is the hallmark of Greek mythology as it is of Egyptian religion and virtually all other ancient legends is that these conflicts are seen from the perspective of human beings. In other words, a battle in heaven is reported by humans as if it has significance for humans. Why should this be so? Why should a war between angels and demons have any relevance for the Earth-bound creatures that we are? And why should such conflicts have been reported to human beings who presumably were not around when they occurred?

The war between Tiamat and Marduk took place long before human beings were spawned and even before the Earth was created, in some dim pre-history of the cosmos. All of these creation epics detail events that could not have been witnessed by any human agency. In that case, who is reporting them? To whom? And why?

We have already reported that, according to the old texts, our blood and breath comes from the gods, from some eldritch lab experiment conducted by the victorious younger deities on the offal of their slain adversaries. So the knowledge of that old combat is programmed into our genes. Let's say we get one set of chromosomes from the old, defeated gods and another set from the younger deities, stirred into a primordial soup formed from the Earth's oxygen, hydrogen, nitrogen and carbon. The knowledge of this ancestry sits embedded deep within the helix like Jung's ancient racial memory: a silent, whispering element of consciousness that hums in the background of our genetic inheritance. It's the elephant in the room, or maybe the eight-hundred-pound gorilla if we credit Darwin (and we do). Whole libraries have been written about virtually every other aspect of biblical lore, about angels and demons, about the Flood, about Adam and Eve, about sin and redemption, but nothing about the first battle, the celestial conflict, the war in heaven. Academics shrug it off, just another fairy tale. Poets like it; Milton enshrined the conflict and Blake teased around its edges. They knew something was there, something vitally important towards an understanding of the human race and of the reality within which the human race must survive. The Manichaeans knew it; the Gnostics knew it. They knew that this conflict rages still.

And the Greeks knew it. And the Greeks knew why it was important for us to remember it, to realize it, and to take sides. That is why they gave us Prometheus.

▼   ▼   ▼

As we saw in Chapter Two, Prometheus was one of the Titans, one of that angry group of godlings that opposed the regime of Zeus and the Olympic gods. Just as importantly, Prometheus was also the god credited with the creation of human beings.

This is an ancient legend, ancient even during the flowering of Greek culture and literature in the eighth century BCE. It has so many parallels with the Sumerian and Babylonian legends as to represent an understanding that transcends any attempt to denigrate the accounts of ancient peoples as mere myths, superstitions, or self-deceptions. The inextricable connection between a conflict in the heavens and the creation of humanity is pronounced in the story of Prometheus. This mythologem has such power that it inspired millennia of poets, scholars, and philosophers down to the present time. If the texts from Sumer and Babylon are sparse in details, the legends of Prometheus are elaborate, articulated, and rich in all manner of plot, sub-plot, characters, and emotive force.

Prometheus is a liminal figure. He occupies a space somewhere between the Titans and the Olympians; although he was a Titan he never engaged in direct battle with Zeus but remained on the sidelines. He not only created the human race, but he stole fire from the Olympians to give to the humans he had created. For that crime he was chained to a rock on a mountain in the Caucasus and every day an eagle would eat his liver. Every night the liver would grow back, and then the eagle would return and so on in an endless series until finally the Greek hero Herakles came and killed the eagle, thus liberating Prometheus.

The idea that Prometheus was responsible for creating humanity was reprised in the nineteenth century with the publication of Mary Shelley's *Frankenstein*. What many may not remember is that the original subtitle to that famous novel is "The Modern Prometheus." The Prometheus in question is not the monster, of course, but Dr. Victor Frankenstein who created the monster: i.e., who created life. Mary Shelley saw modern science usurping the abilities and the ambitions of the ancient Greek Prometheus, assembling the bits and pieces of nerves, limbs, and muscles from an array of dead tissue and combining them to form a living creature in the likeness of a man. Her novel is replete with references to Darwin, and to the medieval alchemists and magicians Agrippa and Paracelsus who themselves were credited with creating homunculi: miniature human beings, like automatons, formed from discarded offal and slime, overheated in glass alembics buried in dung heaps under the solstitial sun.

The creation of humanoid creatures is a venerable tradition among the Jews, with stories about the manufacture of a *golem* (literally "unshaped form" or "raw material") dating to at least the eleventh century CE and the most famous case that of Rabbi Loew of Prague in the sixteenth century: a golem that was created to defend the Jewish ghetto. Even Adam himself was considered a kind of golem by the Rabbis who composed the Talmud and who said that

Adam was created from mud and dust and was thus the prototype of the golems to follow. The difference was that Adam was created by God, whereas the golems of Jewish tradition were created by human beings and therefore did not have the divine spark that was present in Adam. One of the strange characteristics of the Jewish golem was its literalness: it took orders and carried them out, but to the letter. They were incapable of independent thought. Thus, the golem was the prototype of the robot and the computer.

Prometheus, as a Titan, possessed supernatural attributes and was able to create free-thinking human beings just as his Mesopotamian counterpart, Marduk, who could be considered a Sumerian Titan. In the case of Frankenstein, however, we have a human scientist who relegates to himself the capability of doing the same since concepts like divinity, godhead, and the supernatural can have no place in science. Therefore Dr. Frankenstein did not find his abilities circumscribed by Abrahamic ideas about the breath of God animating the lifeless lump of clay that was Adam. To Frankenstein, the creation of life was the proper domain of science and not religion or supernaturalism. All you needed was a collection of the right raw materials and an animating spark or what Dr. Frankenstein called the secret of vitality: something akin to galvanism or electromagnetism, according to the novel.

Appropriately enough, Frankenstein performed his experiments at the University of Ingolstadt in Bavaria. This is the same university at which Adam Weishaupt—the founder of the *Illuminaten Orden*—taught canon law.

Weishaupt's Illuminati—founded on May 1, 1776, about the same time that Shelley's fictional Dr. Frankenstein worked at Ingolstadt—represented the resistance of science and rationalism against the Church and was a political movement designed to undermine monarchical institutions everywhere, even using Masonic lodges as cells in his network of free-thinkers and revolutionaries. One may say that Weishaupt's goal was the same as Victor Frankenstein's: to create the New Man.

Frankenstein's creation—his "monster"—actually referred to himself as the new Adam; but Frankenstein found him hideous and brutish, and rejected him outright just as many believe the Creator abandoned his creation: humanity itself. In the light of modern UFO experiencers and theorists who claim that the aliens are creating hybrids—part human, part alien creatures formed by manipulating human DNA—the stories of golems, homunculi and Frankenstein's monster are particularly relevant. These are modern interpretations of the oldest texts known to humanity, those concerned with the war in heaven, the creation of the cosmos, the fashioning of human beings from dust and divine essences, and the abandonment of creation by a distant pantheon of all-powerful, extraterrestrial beings. More importantly, these literary creations insist on a paradigm that the modern Ufologists generally ignore: the possibility that human beings are active participants in the "hybrid program," so many Doctors Frankenstein performing hideous experiments on human and animal subjects in order to unlock the secrets of life and to decode the mystery of human existence. Mary Shelley combined all of these elements in her novel: the theme of Prometheus, of Adam, Darwin, and the magicians Agrippa, Paracelsus and Albertus Magnus, all in the swirl of revolutionary thought that washed up against the walls of the University of Ingolstadt in the eighteenth century.

Frankenstein's monster turned on its creator. Did Adam turn on God?

Nowhere in the Mesopotamian literature do we read of human beings turning against the gods of the Sumerian or Babylonian pantheons. Nor is this motif apparent in the Egyptian literature, either. The Aztecs, Incas, Mayas . . . all were subservient to the gods who demanded greater and greater sacrifices of blood. Prometheus, the champion of the human race, turned against the gods of Olympus but he was a Titan and a descendant of the gods. In the Bible, while humans may disobey God's commands or attempt to storm heaven with a Tower of Babel, there is no open revolt against God. Humans may ignore God, break God's laws, indulge in licentiousness and all sorts of actions considered contrary to God's commandments, but there was no unified group of human beings who deliberately and consciously defied God.

And then there was Jesus.

We might speculate a little on an alternate significance of this story. Jesus represents—for Christians—the incarnation of God among human beings. In other words, it is God entering the stream of history, entering the flow of his own Creation. An astounding concept, if one thinks about it.

But, then, what happens?

Human beings pounce on this incarnated God, arrest him, beat him, torture him, and then crucify him. Humans finally had a God close to hand, after all those centuries of having God as an absent Father somewhere in the heavens, silent and remote. Here was a God *on Earth*, within easy *reach*.

So they killed him.

▼   ▼   ▼

Christian history for the first thousand years or so was rife with political intrigue. Christians met in secret, in cemeteries and catacombs, under Roman rule until Constantine finally accepted Christianity as a state religion in the fourth century CE. Then followed a period of warring sects and denominations, and eventually the split between the Eastern churches and the Western churches, followed by the Protestant Reformation and more splitting. And the Inquisition. And the Crusades.

By about the time of the Middle Ages a kind of underground movement was taking shape, formed by Catholic priests who no longer believed in the Church or its teachings and who harbored a tremendous hatred of the institution. This manifested in what is known as the Black Mass: ceremonies designed to profane the rituals of the Church and which incorporated sexuality as one of its "sacraments." This was especially true in the celebrated case of the Marquise de Montespan of seventeenth century

France, whose naked body served as the altar in a Black Mass: a historical account that is verified by contemporary observers. Other types of Black Mass were used as occult rituals to cause death to enemies or for other purposes. This is probably as close as one can come to a revolt of human beings against the gods.

For all of its theater, however, even this was nothing more than a replay of Prometheus and his revolt against Olympus. It was the ancient myth, reworked to fit a different culture and time, but the original myth nonetheless: a confused mess of sexuality, violence, spirituality, and the ever-present whiff of alien contact whether as an invocation of the Devil or the desperate urge to kill God and use his blood to create a new life.

The abductees know this story, know it in their bones. No matter how we identify or characterize their experiences—using psychology, or religion, or simply taking them at their word—the abduction is carried out by acolytes of the Unknown God, the Alien God, and the experience is seen through a scrim of sexual manipulation and psychic dislocation. The abductees become an altar in a new Black Mass as the alien forces push their way into their consciousness, impatient and insistent, using whatever sublimated or repressed material they can find buried in their memories or fantasies. And as they work their way towards us, we work our way towards them. As "modern Prometheus" we manipulate matter and energy, causing massive explosions, pandemics, biological clones, hulking robots, vast seas of climate change . . . all the old dreams of the alchemists and the magicians. What has science brought us, after all, but new and better ways of achieving ancient dreams? When, one wonders, will we dare to dream new dreams?

Until we do, the Earth is a hot LZ. LZ is the old Vietnam War era abbreviation for "landing zone," the spot where a helicopter will land. A "hot LZ" is one under fire from the enemy.

Aliens, gods, spirits, ghosts . . . they are taking their lives into their hands landing here.

Or is it part of the original plan, to keep us at each other's throats? When race is not enough to separate us, it's religion. When religion is not enough, it will be economics, or access to clean water and air, or a vaccine to counteract a lethal disease. If you were intent on domination wouldn't you want to see your enemy fragmented and fighting each other? And hasn't this internecine conflict turned global, more deadly than ever before in the history of the world? If you are wondering why the saucers haven't landed on the White House lawn—or in Tian An Men Square, or Red Square, or the Champs Elysée—isn't it because they are waiting for us to do all the heavy lifting first? When the dust and smoke clears, when the last bodies have been burned or buried, then they will show up and when they do it will go unnoticed and uncelebrated.

▼   ▼   ▼

What we have seen so far in this volume is a history of religious attitudes towards a handful of experiences that are shared across cultures and eras. This is Ufology as understood by the ancients. This is von Däniken territory, Zecharia Sitchin territory, and the domain of the ancient alien theorists, but we have taken a different approach. We see a pattern forming out of all this varied data, as sure as we test high on the spectrum. We don't need to deconstruct the Nazca lines or the Great Pyramid to arrive at the same conclusion. The proof is in the texts themselves, if you know where to look. If you can see the forest for the trees.

We are the living evidence of contact. We carry that evidence with us, in our genes possibly, but certainly in our consciousness, in our minds. Contact is part of our heritage, it is what has kept us going, it is the reason we have progressed so much in the last ten thousand years (more than in the previous two hundred thousand years). We are working our way towards a comprehensive understanding of this contact, but we are doing it as sleepwalkers. It forms our motivation, our reason for working and striving in the sciences and in every other human endeavor, but we are unconscious of the real purpose behind what we do.

As the twentieth century dawned, however, it gradually became more clear. The ferment of science, politics and religion out of which that century exploded demonstrates how close we came to a real understanding of the nature of reality and of our role within the world. It is encapsulated in the Phenomenon, which by the mid-twentieth century was banging down our door.

The Phenomenon: you can see its outline taking shape in the preceding chapters. It is the UFO phenomenon, to be sure, but that is a box too small to contain it in all its glory especially as no two experts can agree on what it is. The Phenomenon extends to virtually every aspect of modern experience, influencing religion (of course) but also science, technology, medicine, the arts, and the media. There is no denying that the Phenomenon is dangerous: some of the best minds of our generation have been ruined by too close proximity to it. Reputations have been destroyed. People deceived.

Politics is not immune to its influence. The roster of politicians, generals, and other thought leaders who have had direct experience of it is long and distinguished; their names have been published, their observations credited.

It used to be believed that it was dangerous to waken a sleepwalker. In fact, the opposite is true: it is necessary to rouse the sleepwalker—gently—so that he or she does not harm themselves. It is time now to awaken the sleepwalker in all of us.

"And do you dream?" said the demon.

— Mary Shelley, *Frankenstein*

This is a painting from 1623 by Dirck van Baburen depicting Prometheus being chained. Notice the presence of Mercury with his caduceus: the twin serpents around a central shaft. This motif will be investigated in greater detail in Book Three.

Prometheus is being chained because he committed the sin of bringing the gift of fire to humanity. He will be chained to a rock, his liver eaten out of him by an eagle every day which will grow back at night, only to be eaten again the next day until he is rescued by Herakles. The juxtaposition of Mercury - the messenger of the gods - with the imprisonment of Prometheus is very revealing, as we will see.

Endnotes

INTRODUCTION

1   *Janes Defense Weekly*, June, 1997, as cited in Goliszek, Andrew, *In the Name of Science: A History of Secret Programs, Medical Research, and Human Experimentation*, St. Martin's Press, NY, 2003, p. 261.

2   STEM is the acronym for Science, Technology, Engineering and Math. STEAM includes the Arts.

CHAPTER ONE

1   Aimé Michel, "The Problem of Non-Contact," *The Humanoids*, Charles Bowen, ed. (London: Futura Publications, 1977), p.255.

2   Thorkild Jacobsen, *The Treasures of Darkness: A History of Mesopotamian Religion*, (New Haven, CT: Yale University Press, 1976), p. 3.

3   Hans Jonas, *The Gnostic Religion: The Message of the Alien God and the Beginnings of Christianity* (Boston: Beacon Press, 1958, 1963), p. 50.

4   Jonas (1963) p. 42.

5   Ibid., p. 43.

6   Jeffrey D. Kripal, *Authors of the Impossible: The Paranormal and the Sacred* (Chicago: University of Chicago Press, 2010).

7   Thomas van Flandern, *Dark Matter, Missing Planets and New Comets: Paradoxes Resolved, Origins Illuminated* (Berkeley: North Atlantic Books, 1999).

8   Thomas van Flandern (1999), p. xvi-xvii.

9   Samuel Noak Kramer, *History Begins at Sumer*, (Philadelphia: University of Pennsylvania Press, 1956, 1981).

10   Giorgio de Santillana, and Hertha von Dechend, *Hamlet's Mill: An essay investigating the origins of human knowledge and its transmission through myth* (Boston: David R. Godine, 1977).

11   L. A. Waddell, *Egyptian Civilization: Its Sumerian Origin & Real Chronology and Sumerian Origin of Egyptian Hieroglyphs* (London: Luzac & Co., 1930). Waddell was criticized for the same reasons as Sitchin: that he mistranslated and misinterpreted the Sumerian language, and that he took many fanciful tales to be literally true. To his credit, Waddell was an acknowledged expert in Tibetan language and culture, was a part of Younghusband's 1904 invasion of Tibet, and was in China during the Boxer Rebellion. However, his views on Aryan racial supremacy and on Sumer as the Aryan ur-civilization that diffused throughout the world, jump-starting all of the world's other civilizations and religions, marginalized his work among scholars, linguists and historians.

12   Robert K. G. Temple, *The Sirius Mystery* (New York: St. Martin's Press, 1976).

13   Peter Levenda, *The Tantric Alchemist: Thomas Vaughan and the Indian Tantric Tradition* (Lake Worth, FL: Ibis Press, 2015).

14   For a recent overview of these theories see Christopher Knowles, *Our Gods Wear Spandex: The Secret History of Comic Book Heroes* (San Francisco: Weiser Press, 2007).

15   The terms *kaššāpu* or *kaššāptu* (male and female, respectively) are usually translated as "witch" with the understanding that this term means a magic practitioner who is intent on causing evil: sickness, death, etc., but using the same powers as those employed by the exorcist and the "state-approved" magic practitioner. See Tzvi Abusch, *Mesopotamian Witchcraft: Towards a History and Understanding of Babylonian Witchcraft Beliefs and Literature* (Leiden: Brill, 2002), p. 7.

16   Jonas, pp. 75-80.

17   I am thinking of Jim Marrs, Richard Dolan, Jacques Vallée, and others who straddle both sides of the discussion in their works, with Marrs beginning with the Kennedy assassinations and moving towards Ufological works, and Dolan beginning with Ufology and moving towards political conspiracies, with Vallée firmly in the middle of science and IT on one side and Ufology and a kind of mysticism on the other.

18 Dora Jane Hamblin, "Has the Garden of Eden been located at last?" *Smithsonian* magazine, Volume 18, No. 2 (May 1987), on the work of archaeologist Dr. Juris Zarins.

19 We will explore this theme in greater detail in Book Three where we examine the structure and origin of the genetic code, its mathematics, and its irruption into the pre-scientific mind in Asia and Africa through various divination systems as well as the ubiquitous game of chess.

20 See for instance Dr. Kenneth Ring, "Near-Death and UFO Encounters as Shamanic Initiations," *ReVision*, Vol. 11, No. 3 (Winter 1989); and Lorraine Davis, "A Comparison of UFO and Near-Death Experiences as Vehicles for The Evolution Of Human Consciousness," *Journal of Near-Death Studies*, 6 (4) (Summer 1988), pp. 240–257, incidentally a text that cites Arthur Young as a reference (see Book Two).

21 We will learn much more about Puthoff, Targ, Swann and SRI in Book Two.

22 See for instance Mikael Rothstein, "UFO beliefs as syncretistic components," in Christopher Partridge, ed., *UFO Religions* (New York: Routledge, 2003), the section on Scientology on pp. 263–265.

23 Jacques Vallée, *Forbidden Science: Journals 1957–1969* (Berkeley: North Atlantic Books, 1992), p. 33 and p. 228.

24 This phrase occurs in handwritten notes attached to a Defense Intelligence Agency memo dated January 26, 1995, ORD-0143-95 to the Associate Deputy Director for Military Affairs, Office of Military Affairs, DO (CIA-RDP96-00791R000100140001-2). The line reads "SAP - LIMDIS - ? Giggle Factor." SAP stands for "Scientific Advisory Personnel," and LIMDIS stands for "Limited Distribution," a classification level.

25 See for instance the declassified 1995 US Department of Defense, General Defense Intelligence Program memorandum S-095/GS from Letitia A. Long, Director, DMI Staff, with its attached STAR GATE retrospective review, p. 2: "Under the division of labor in the Army GRILL FLAME project, MIRADCOM had responsibility for developing a Remote Perturbation experimental program . . ." STAR GATE and GRILL FLAME were code words for military remote viewing projects; MIRADCOM stands for the "Missile Research and Development Command" of the US Army.

26 Of possible relevance is the declassification of some remote viewing tasking that took place during the first Gulf War, which resulted in

two "hits," two "partial hits," one "probable hit," one "apparent miss" and one "miss." See CIA-RDP96-00789R003600350002-3, Project Feedback Sheet, Project 91002, dated 910206. Some of this tasking had to do with Saddam Hussein's location at the time (1991) as well as predicted terror attacks.

27   Victor Turner, *The Ritual Process: Structure and Anti-Structure* (Aldine, Chicago, 1969).

28   Mircea Eliade, *Shamanism: Archaic Techniques of Ecstasy* (New York: Pantheon Books, 1964).

29   An example might be Daniel C. Noel, *The Soul of Shamanism: Western Fantasies, Imaginal Realities*, (New York: Continuum, 1997), and the essays on Eliade in Christian K. Wedemeyer and Wendy Doniger, eds., *Hermeneutics, Politics, and the History of Religions: The Contested Legacies of Joachim Wach and Mircea Eliade* (New York: Oxford University Press, 2010).

30   R. D. Laing, *The Politics of Experience* (New York: Pantheon Books, 1967).

31   The *Picatrix* was not translated into Latin until the thirteenth century, and it is likely that this is a version of an earlier, tenth-century work originally written in Arabic.

32   Gösta Hedegård, *Liber Iuratus Honorii: A Critical Edition of the Latin Version of the Sworn Book of Honorious* (Stockholm: Almqvist & Wiksell International, 2002), p.36.

CHAPTER TWO

1   Kurt Seligmann, *The History of Magic and the Occult* (New York: Random House, (1948, 1997)), p. 1.

2   In quantum physics, the "many worlds interpretation" (or MWI) holds that many worlds exist in order to accommodate every possible outcome of a random event. This is also known as the Everett Postulate, which claims that the universe is composed of many superimposed states. We will discuss this in greater detail in Book Three. In a way, these quantum theories can be understood as "myths": common language examples used to explain principles that can only be understood using higher mathematics and advanced physics. See the work by astrophysicist John Gribbins, *Schrodinger's Kittens and the Search for Reality*, (London: Phoenix, 1995) for an elaboration of this idea that vernacular descriptions of quantum effects are not true in

a literal sense but act as myths, as ways of explaining concepts that otherwise resist language. The truth is contained in the equations, and not in the explanations of them: an idea that has profound implications for the study of the Phenomenon and which may help us interpret Sumerian and other texts from that point of view.

3   This was described by authors such as James Frazer as the *hieros gamos* or sacred marriage: a ritual known in many places around the world. An almost identical ritual is performed each year to this day in Indonesia, between the sultan of Yogyakarta and a mysterious goddess, notwithstanding the fact that the sultan is a Muslim in a Muslim-majority country. See Peter Levenda, *Tantric Temples* (Lake Worth, FL: Ibis Press, 2011) for a discussion of this ritual in its Indonesian context.

4   That the God of the Israelites did have a consort originally—known as Asherah—is now acknowledged, but was expurgated from the Bible by the compilers of the Pentateuch.

5   Alan F. Segal, *Two Powers in Heaven: Early Rabbinic Reports about Christianity and Gnosticism* (Leiden: Brill, 2002).

6   See, for instance, Karen Sonik, "The Tablet of Destinies and the Transmission of Power in Enūma eliš," in Gernot Wilhelm, ed., *Organization, Representation and Symbols of Power in the Ancient Near East* (Winona Lake, IN: Eisenbrauns, 2012), pp. 387–395.

7   Alexander Wendt and Raymond Duvall, "Sovereignty and the UFO," *Political Theory*, Volume 36, Number 4 (August 2008), pp. 607-633.

8   Wendt and Duvall, p. 608.

9   See for instance the Spanish word *real* which can mean either "royal" (*el camino real*) or "real" (as in *realidad* or *lo real*).

10   Wendt and Duvall, p. 610.

11   Michael S. Heiser, *The Unseen Realm: Recovering the Supernatural Worldview of the Bible* (Bellingham, WA: Lexham Press, 2015), pp. 106–107.

12   According to several sources, the name of the first of the seven apkallu was Uanna, which may be the original form of Oannes, transliterated into Greek.

13   R. Campbell Thompson, *The Epic of Gilgamesh: Complete Academic Translation* (London: Luzac & Co. 1928).

14   As described in Andrew R. George, "The Gilgameš Epic at Ugarit," *Acta Orientalis* 25 (2007) 237-254.

CHAPTER THREE

1   This statement should not escape the attention of a close reader, for in Latin it is usually translated as *Terribilis est locus iste* and as such is found over the entrance to the famous Church of Saint Mary Magdalene at Rennes-le-Chateau, the center of much of the controversy surrounding the book *Holy Blood, Holy Grail* (Michael Baigent, et al.) and the Dan Brown book and film, *The DaVinci Code*. It is often—mistakenly—translated as "This place is terrible," as if it were a warning. However, this quotation is found engraved or carved in the buildings of many churches worldwide including more than twenty in Italy alone. It is taken from this chapter of Genesis and is used to denote any spot that is sacred to God.

2   Ahmad H. Sakr, *Al-Jinn* (Lombard, IL: Foundation for Islamic Knowledge, 1994).

3   It may be useful to remind the reader here that the motif of being "wounded in the thigh" has an honorable pedigree in the tales surrounding the Holy Grail, the Knights of the Round Table, and the Fisher King. It was a euphemism for being wounded in the genitalia, either castration or a direct wound to the penis, thus rendering the man impotent. Further, the instrument used for the Egyptian "opening of the mouth" ritual mentioned above was "the thigh of Set" (which was also the name of the Big Dipper constellation).

4   Kathryn Pocalyco, "Jacob's Hip: An Anatomical Inquiry Into Genesis Chapter 32," *Glossolalia*, 3.1 (December): 105113.

5   Peter Levenda, *Stairway to Heaven* (New York: Continuum, 2008); and Levenda, *The Secret Temple* (New York: Continuum, 2009).

6   Tradition has it that Genesis, the Song of Solomon, and the Book of Ezekiel were the three texts of the Tanakh that were surrounded by Jewish prohibition and taboo. Of the three, Ezekiel was considered the most dangerous.

CHAPTER FOUR

1   From Walter Lehmann, *Colloquies and Christian Doctrine*, (Stuttgart, 1949), cited in Miguel León-Portilla, *Aztec Thought and Culture* (Norman, OK: University of Oklahoma Press, 1963, 1990), p. 63.

2   Miguel León-Portilla, *Aztec Thought and Culture* (Norman, OK: University of Oklahoma Press, 1963, 1990), p. 75.

3   León-Portilla (1990), p. 123.

4   With the exception, perhaps, of Tibetan Buddhism where this practice has been retained in several ritual and meditative forms, such as *dcod.* (see below).

5   See the writings, for instance, of Paul Manansala on this subject, which posits an "Atlantis" in the Pacific for which he considers "Aztlan" a cognate term.

6   There is some similarity to the story of the Exodus. In both cases, people decide to leave their residence on a long migration to find a more suitable homeland, guided by a god speaking through a human agent. They do not leave—according to some sources—due to having been thrown out or because of natural disasters, but because there is a religious conflict between them and the rulers of Aztlan. See Michael Pina, "The Archaic, Historical and Mythicized Dimensions of Aztlan" in Rudolfo A. Anaya, Francisco Lomeli, eds., *Aztlan: Essays on the Chicano Homeland* (Albuquerque: University of New Mexico Press, 1981), pp. 14-48 for a discussion of the problem.

7   John K. Grandy, "The Three Neurogenetic Phases of Human Consciousness," *Journal of Conscious Evolution*, Issue 9 (2013), pp. 1-24; "The Neurogenetic Substructures of Human Consciousness," *Essays in Philosophy*, Volume 15, Issue 2, Article 3 (2014); "DNA Consciousness: From Theory to Science," in Ingrid Frederikson, ed., *The Mysteries of Consciousness: Essays on Spacetime, Evolution and Well-Being* (Jefferson, NC: McFarland & Co. Inc., 2015), pp. 116-148.

8   Todd E. Feinberg, Jon Mallatt, "The evolutionary and genetic origins of consciousness in the Cambrian Period over 500 million years ago," *Frontiers in Psychology*, Volume 4, Article 667 (October 2013).

9   Vladimir I. shCherbak and Maxim A. Makukov, "The 'WOW! Signal' of the terrestrial genetic code," in *Icarus* 224 (2013) 228-242.

CHAPTER FIVE

1   R.T. Rundle Clark, *Myth and Symbol in Ancient Egypt* (London: Thames & Hudson, 1959, 1991), pp. 32-33.

2   See the excellent approach to the Phenomenon in Whitley Strieber and Jeffrey D. Kripal, *The Super Natural: A New Vision of the Unexplained* (New York: Penguin, 2016), where this aspect is discussed in more detail.

3   See Book Two, Chapter Four.

4   *Blade Runner* (Warner Bros., 1982).

5   Vincent Arieh Tobin, "Creation Myths," *The Ancient Gods Speak: A Guide to Egyptian Religion*, Donald B. Redford, ed., (New York: Oxford University Press, 2002), p. 246.

6   Toby Wilkinson, "Before the Pyramids: Early Developments in Egyptian Royal Funerary Ideology," *Egypt at its Origins. Studies in Memory of Barbara Adams*, S. Hendrickx, Friedman, Ciałowicz, Chłodniki, eds. (Leuven: Peeters, 2004) p. 1141. Emphasis added.

7   Andrey O. Bolshakov, "Ka-Chapel," *The Ancient Gods Speak: A Guide to Egyptian Religion*, Donald B. Redford, ed. (Oxford: Oxford University Press 2002), p. 182.

8   Again, there is a degree of ambiguity when dealing with the concepts of space and time in the Creation epics of the Jews and the Egyptians. There were no "days" as such before the creation of the sun which did not take place, Genesis tells us, until the fourth "day"; "day," therefore, refers to a division of time that is not equivalent to an Earth day. It may be instructive to view our traditional seven-day week as a kind of metaphor of the astronomical periods of time in which the universe—and, eventually, our solar system—was created in the aftermath of the Big Bang. That there are calendars in the world that do not use a seven-day week may come as a surprise to many readers.

9   Gertie Englund, "Offerings," *The Ancient Gods Speak*, Donald B. Redford, ed., p. 286.

10 This topic is covered extensively in Louis V. Žabkar, *A Study of the Ba Concept in Ancient Egyptian Texts* (Chicago: University of Chicago Press, 1968).

11 Žabkar, p. 95.

12 Whether this is an avoidance of death itself or only of pain or confrontation that could lead to pain is an existential question that is impossible to answer. Humans have a fear of death that is unrelated to pain: it is a fear of loss of identity in some cases, or a fear inculcated by religious teachings that threaten an unpleasant afterlife.

13 Such as the discovery of what may be America's oldest city, the Caral complex in Supe Valley, Peru. This complex dates to the same time frame as the Djoser Step Pyramid—roughly 2800 to 2600 BCE—and, indeed, contains pyramids as part of its construction.

14　This point has also been suggested by Robert Temple in his book *Egyptian Dawn* (London: Random House, 2010).

15　See http://www.dailygrail.com/Hidden-History/2015/4/Game-Groans-Egyptologist-Zahi-Hawass-Goes-Meltdown-During-Debate-Graham-Hancoc, accessed June 1, 2015.

16　Robert Temple, *The Sirius Mystery*, (London: Arrow Books, 1999), pp. 61-69.

17　Giorgio de Santillana and Hertha von Dechend, *Hamlet's Mill: An Essay Investigating the Origins of Human Knowledge and its Transmission through Myth* (Boston: David R. Godine, 1977), p. xii.

18　All we know for sure is that the mastabas were oriented along the north-south axis, as were the later pyramids. This may indicate the beginning of a sensitivity to the northern quadrant as the place of immortality. In what may be an unrelated case, the Yezidi of northern Iraq also bury their dead facing the north; in Islam, the burials normally face Mecca.

19　See the works of Emile Durkheim and Max Weber who took a sociological approach to the decline of traditional religion and the rise of more secular forms. That communism was considered a "secular religion" was a common idea in the 1930s, with the Party as the church and the *Communist Manifesto* as the scripture.

20　Ironically, although the Russian architect Aleksey Shchusev was the designer of record, it was a Jew—Isidore Frantsuz—who was the real designer of the pyramid-like tomb. This information was suppressed at the time due to the rampant anti-Semitism of the Soviet regime. This is reminiscent of the belief, current in some circles, that the Jews built the original pyramids.

21　Cited in Robert Markley, *Dying Planet: Mars in Science and the Imagination* (Durham: Duke University Press, 2005), pp. 134–135. This was certainly a surreal conversation, seen in hindsight: the author of *The War of the Worlds* and *The Time Machine* interviewing the father of the Russian Revolution about aliens!

22　Michael G. Smith, *Rockets and Revolution: A Cultural History of Early Spaceflight*, (Lincoln: University of Nebraska Press, 2014), p. 110.

23　Ibid., pp. 294-295.

24　See Robert Temple, *Egyptian Dawn*, 2010, p. 60, where this phenomenon is discussed with its implications for mummification.

25   The American anthropologist, W. Y. Evans-Wentz, drew that comparison immediately, which is why he called the first-ever English translation of the Tibetan document the "Tibetan Book of the Dead" as a reference to its similarity to the Egyptian Book of the Dead. *The Tibetan Book of the Dead* (New York: Oxford University Press, 1927, 1969).

26   This characterization is from Evans-Wentz (1927) and is buttressed by the "Psychological Commentary" by Carl G. Jung that accompanies the 1957 and later editions. The Evans-Wentz version has been criticized by later scholars as showing undue influence from Theosophical ideas as he had been a member of the Theosophical Society since 1901. However, the translations were not his but were carried out by two Tibetan lamas.

27   Anthony P. Sakovich, "Explaining the Shafts in Khufu's Pyramid at Giza," *Journal of the American Research Center in Egypt*, Volume 42 (2005-2006), pp. 1-12.

28   Robert Temple, *Egyptian Dawn* (2010), pp. 43-80.

29   William F. Morain, *Mysteries of the Hopewell: Astronomers, Geometers, and Magicians of the Eastern Woodlands* (Akron: University of Akron Press, 2000), pp. 143-160.

CHAPTER SIX

1   Isabelle Robinet, "Visualization and Ecstatic Flight in Shangqing Taoism," *Taoist Meditation and Longevity Techniques*, Livia Kohn, ed., (Ann Arbor: The University of Michigan, 1989), pp. 181-182.

2   Siegbert Hummel, "La scrittura dei Na-khi," *Tracce d'Egitto In Eurasia*, Guido Vogliotti, ed., (Torino: Ananke, 1997), pp. 122-135.

3   Stephen Oppenheimer, *Eden in the East: The Drowned Continent of Southeast Asia*, (London: Weidenfeld and Nicolson, 1998).

4   Toby Wilkinson, *Genesis of the Pharaohs: Dramatic New Discoveries Rewrite the Origins of Ancient Egypt* (London: Thames & Hudson, 2003).

5   Robert M. Schoch, *Voyages of the Pyramid Builders: The True Origins of the Pyramids from Lost Egypt to Ancient America*, (New York: Putnam, 2003).

6   Schoch (2003), pp. 75-76.

7   Ibid., p. 270.

8   Jennifer Elisabeth Hellum, *The Presence of Myth in the Pyramid Texts*, University of Toronto, (Unpublished thesis, 2001), p. 12.

9   Philip Rawson and Laszlo Legeza, *Tao: The Chinese Philosophy of Time and Change* (London: Thames & Hudson, 1973/1979), p. 31.

10   See for instance Sarah M. Nelson, et al., "Archaeoastronomical Evidence for *Wuism* at the Hongshan Site of Niuheliang," *Journal of East Asian Material Culture* (April 19, 2006).

11   Edward H. Schafer, *Pacing the Void: T'ang Approaches to the Stars* (Los Angeles: Floating World Editions, 2005), p. 234.

12   Schafer, p. 239.

13   Anne Birrell, *Chinese Mythology: An Introduction* (Baltimore: The Johns Hopkins University Press, 1993), p. 132.

14   Joseph Eddy Fontenrose, *Python: A Study of Delphic Myth and its Origins* (Berkeley: University of California Press, 1959, 1980), p. 494.

15   Schafer, p. 239.

16   Isabelle Robinet, (1989), p. 179.

17   Isabelle Robinet, "Visualization and Ecstatic Flight in Shangqing Taoism," *Taoist Meditation and Longevity Techniques*, Livia Kohn, ed., (Ann Arbor: University of Michigan, 1989), p. 184.

18   Ibid., p. 178.

19   Ibid., p.177.

20   Ibid., p.176.

21   Nicole Casal Moore, "Dark matter powered the first stars, physicists speculate," University of Michigan website, Dec. 14, 2007, http://ns.umich.edu/new/releases/6230, last accessed June 15, 2015.

22   Robert Temple, *Egyptian Dawn: Exposing the Real Truth behind Ancient Egypt* (London: Random House, 2011), pp. 370-434.

23   Sarah M. Nelson, et al. (2006), p. 2.

CHAPTER SEVEN

1   Jeffrey Hopkins, "Introduction," Tenzin Gyatso, the Fourteenth Dalai Lama, *Kalachakra Tantra: Rite of Initiation* (Boston: Wisdom Publications, 1985, 1999), p. 65.

2   C. G. Jung, *Flying Saucers: A Modern Myth of Things Seen in the Sky* (New York: Routledge, 1959, 2002), p. 133

3   H. S. Mukanda, et al., "A Critical Study of the Work 'Vymanika Shastra,'" *Scientific Opinion* (1974), pp. 5-12.

4   This theory is hotly contested by some who claim the Aryan Migration Theory is at heart a colonialist perspective that does not reflect the possibility that Indian culture and religion developed independently on the sub-continent. While this is not the mainstream position it is still worth mentioning here. See the works of Michael Witzel on this subject.

5   Wash Edward Hale, *Asura in Early Vedic Religion* (Delhi: Motilal Banarsidass, 1986), pp. 183-193.

6   Ananda K. Coomaraswamy, "Angel and Titan: An Essay in Vedic Ontology," *Journal of the American Oriental Society*, Vol. 55, No. 4, (Dec. 1935), pp. 373-419.

7   W.B. Henning, "The Book of Giants," *Bulletin of the School of Oriental and African Studies*, University of London, Vol. XI, Part 1, (1943), pp. 52-74.

8   Satapatha Brahmana, Eighth Adhyaya, First Brahmana, verses 1-6, in Julius Eggeling, *Satapatha Brahmana Part 1* (Oxford: The Clarendon Press, 1882), pp. 216-218.

9   In another version of the same story, Manu was accompanied on his ship by his family and by the Seven Sages. The Seven Sages are represented in Indian religion by the seven stars of the Dipper and, indeed, the mountain on which Manu's ship runs aground is called the Northern Mountain: another possible allusion to the Dipper and the polestar.

10 An ancient Latin palindrome in the form of a magic square found inscribed on a wall in Pompeii and reproduced in many places as a charm or talisman. It resists translation, but *"opera rotas"* may mean "work of the wheel."

11 T. B. Karunaratne, "The Buddhist Wheel Symbol," *The Wheel Publication* No. 137/138 (Kandy (Sri Lanka): Buddhist Publication Society, 1969, 2008), p. 3. The English transliteration is from the Pali version; a transliteration from the Sanskrit would render *cakka* as *cakra* and *cakkavatti* as *cakravarti*.

12 T. B. Karunaratne, (1969/2008), p. 4.

ENDNOTES

13   John Newman, "Vajrayoga in the Kalacakra Tantra," *Tantra in Practice*, David Gordon White, ed. (Princeton: Princeton University Press, 2000), p. 588.

14   The *Sumaṇ galavilāsinī* of Buddhaghosa, Part II, p. 617 ff. as quoted in Karunaratne, p. 9.

15   Jacques Vallée, *The Invisible College* (New York: E. P. Dutton & Co., 1975), p. 197.

16   Vallée, (1975), pp. 174-175.

17   John Newman, (2000), p. 591.

18   Ibid., p. 592.

CHAPTER EIGHT

1   Max Muller, *Comparative Mythology: An essay* (London: Routledge & Sons, 1909), p. 14.

2   Andrija Puharich, *Uri: A journal of the mystery of Uri Geller* (New York: Anchor Press, 1974), p. 126.

3   Max Muller (1909), p. 16.

4   The hieroglyphics recorded as having been discovered on the famous Roswell UFO debris bore no obvious relationship to the Greek alphabet; however, symbols on another "I-beam" photo bear resemblance to Greek letters delta, omicron, lambda, sigma, and others. Attempts to translate or interpret the word or words created by these letters vary widely, with one interpretation having the letters spell "eleutheria" or "freedom" in Greek.

5   George Steiner, *Antigones* (Oxford: Clarendon Press, 1986), pp. 124-125.

6   Hans Jonas, *The Gnostic Religion: The Message of the Alien God and the Beginnings of Christianity* (Boston: Beacon Press, 1958, 1963), pp. 42-43.

7   Michael Wise, et. al., "The Book of Giants," 4Q530, 4Q203, *The Dead Sea Scrolls: A New Translation* (San Francisco: HarperCollins, 1996, 2005), p.292.

8   "The Book of Giants," 4Q531 Frag.1, (1996, 2005), p. 291.

9   There is a third possibility, of course, and that is the alien contacts we have experienced in more recent times are not of the forces that seeded

this planet at all but are of other classes of being entirely with no connection to the race that seeded us. In that case, either we share some genetic material with this third race (which may have come from the same original source) or there is no common genetic denominator at all.

10 From this perspective, of course, *all* religions are UFO religions: a theme we will pick up again as we go along in this trilogy.

Index

similarities with China and Egypt, 297, 324

Tantras, 9–10, 341, 343

Vedic period in, 324–325

initial contacts. *See* contact

initiation

in Freemasonry, 34–35

Kalacakra, and UFO sightings and, 348–349, 352–353, 354

in shamanism, 34, 36

Tibetan Buddhist, wheel symbol in, 340

intelligence agents, 237–238

Invisible College, The, 347–348, 350

Isaac and Abraham, 120–121, 179

J

Jackson, Georgianna, li

Jackson, Henry, li

Jacob

contention between Esau and, 120, 122, 131–132

stone anointed by, 125

wounding and limp of, 130, 132–133, 400

wrestling with the angel, 119, 129–132

Jacob's Ladder, 118–125

awesome gate of heaven in, 123, 399

Blake's painting of, 149

experienced at night, 132

other world in, 124

summary of the vision, 118, 122–123

two-way traffic on, 124

Jacobsen, Thorkild, 1, 2

Jesus, 387–388

Jewish tradition. *See also* Bible, the

Ezekiel's Vision prohibitions in, 137, 142, 400

golem in, 384–385

Hellenistic influence on, 105

merkava mysticism in, 116

Sumerian beliefs retained in, 82–83

Tower of Babel in, 83–84

two powers in, 71

jinn, 128

Jonas, Hans, 1, 2

Jung, Carl G., 205, 323, 341

K

ka (spirit double), 215, 218–219, 264, 266

Ka'aba of Mecca, 125

Kalacakra Tantra

initiatory mandala in, 343, 347

nonverbal communication in, 343–344

Shambala in, 354

UFO sightings and initiations, 348–349, 352–353, 354

Vimalaprabhā commentary on, 351–353

wheel symbol in, 336, 340–341

Kali (Indian goddess), 158, 177–178

Kelly, Edward, 54

Kent, Clark, 12–13

Key of Solomon, The, 45, 58

Kim il-Sung, 247

Kramer, Heinrich, 172

Kramer, Samuel Noah, 6

Kripal, Jeffrey J., liv, lvi, 5

Krupp, E. C., 232

Kundalini yoga, 51

L

La Venta pyramid in Tabasco, 195

Laing, R. D., 37–38

Lao Ze (Lao Tse), 293

Layard, Austen Henry, 7, 327

INDEX

ABOUT THE AUTHORS

Tom DeLonge is an award-winning American musician, producer, and director best known as the lead vocalist, guitarist, and songwriter for the platinum-selling rock bands Blink-182 and Angels & Airwaves. His home is San Diego,  California, where he focuses on creating entertainment properties that cross music, books, and film with his company To The Stars... Check out his other multi-media projects at ToTheStars.Media.

▼   ▼   ▼

Peter Levenda has an MA in Religious Studies and Asian Studies from FIU, and speaks a variety of languages (some of them dead). He has appeared in numerous television programs for the History Channel, the  Discovery Channel, National Geographic, and TNT. He has interviewed Nazis, Klansmen, occultists, CIA officers, and Islamic terrorists in the course of his research, and has visited Chinese prisons and military bases; the Palestine Liberation Organization; the former KGB headquarters (Dzherzinsky Square) in Moscow; and once was celebrated with a state dinner in Beijing.